Lecture Notes in Mathematics

For information about Vols. 1–1173 please contact your bookseller or Springer-Verlag.

Vol. 1174: Categories in Continuum Physics, Buffalo 1982. Seminar. Edited by F.W. Lawvere and S.H. Schanuel. V, 126 pages. 1986.

Vol. 1175: K. Mathiak, Valuations of Skew Fields and Projective Hjelmslev Spaces. VII, 116 pages. 1986.

Vol. 1176: R.R. Bruner, J.P. May, J.E. McClure, M. Steinberger, H_∞ Ring Spectra and their Applications. VII, 388 pages. 1986.

Vol. 1177: Representation Theory I. Finite Dimensional Algebras. Proceedings, 1984. Edited by V. Dlab, P. Gabriel and G. Michler. XV, 340 pages. 1986.

Vol. 1178: Representation Theory II. Groups and Orders. Proceedings, 1984. Edited by V. Dlab, P. Gabriel and G. Michler. XV, 370 pages. 1986.

Vol. 1179: Shi J.-Y. The Kazhdan-Lusztig Cells in Certain Affine Weyl Groups. X, 307 pages. 1986.

Vol. 1180: R. Carmona, H. Kesten, J.B. Walsh, École d'Été de Probabilités de Saint-Flour XIV – 1984. Édité par P.L. Hennequin. X, 438 pages. 1986.

Vol. 1181: Buildings and the Geometry of Diagrams, Como 1984. Seminar. Edited by L. Rosati. VII, 277 pages. 1986.

Vol. 1182: S. Shelah, Around Classification Theory of Models. VII, 279 pages. 1986.

Vol. 1183: Algebra, Algebraic Topology and their Interactions. Proceedings, 1983. Edited by J.-E. Roos. XI, 396 pages. 1986.

Vol. 1184: W. Arendt, A. Grabosch, G. Greiner, U. Groh, H.P. Lotz, U. Moustakas, R. Nagel, F. Neubrander, U. Schlotterbeck, One-parameter Semigroups of Positive Operators. Edited by R. Nagel. X, 460 pages. 1986.

Vol. 1185: Group Theory, Beijing 1984. Proceedings. Edited by Tuan H.F. V, 403 pages. 1986.

Vol. 1186: Lyapunov Exponents. Proceedings, 1984. Edited by L. Arnold and V. Wihstutz. VI, 374 pages. 1986.

Vol. 1187: Y. Diers, Categories of Boolean Sheaves of Simple Algebras. VI, 168 pages. 1986.

Vol. 1188: Fonctions de Plusieurs Variables Complexes V. Séminaire, 1979–85. Edité par François Norguet. VI, 306 pages. 1986.

Vol. 1189: J. Lukeš, J. Malý, L. Zajíček, Fine Topology Methods in Real Analysis and Potential Theory. X, 472 pages. 1986.

Vol. 1190: Optimization and Related Fields. Proceedings, 1984. Edited by R. Conti, E. De Giorgi and F. Giannessi. VIII, 419 pages. 1986.

Vol. 1191: A.R. Its, V.Yu. Novokshenov, The Isomonodromic Deformation Method in the Theory of Painlevé Equations. IV, 313 pages. 1986.

Vol. 1192: Equadiff 6. Proceedings, 1985. Edited by J. Vosmansky and M. Zlámal. XXIII, 404 pages. 1986.

Vol. 1193: Geometrical and Statistical Aspects of Probability in Banach Spaces. Proceedings, 1985. Edited by X. Fernique, B. Heinkel, M.B. Marcus and P.A. Meyer. IV, 128 pages. 1986.

Vol. 1194: Complex Analysis and Algebraic Geometry. Proceedings, 1985. Edited by H. Grauert. VI, 235 pages. 1986.

Vol. 1195: J.M. Barbosa, A.G. Colares, Minimal Surfaces in $\mathbb{R}^3$. X, 124 pages. 1986.

Vol. 1196: E. Casas-Alvero, S. Xambó-Descamps, The Enumerative Theory of Conics after Halphen. IX, 130 pages. 1986.

Vol. 1197: Ring Theory. Proceedings, 1985. Edited by F.M.J. van Oystaeyen. V, 231 pages. 1986.

Vol. 1198: Séminaire d'Analyse, P. Lelong – P. Dolbeault – H. Skoda. Seminar 1983/84. X, 260 pages. 1986.

Vol. 1199: Analytic Theory of Continued Fractions II. Proceedings, 1985. Edited by W.J. Thron. VI, 299 pages. 1986.

Vol. 1200: V.D. Milman, G. Schechtman, Asymptotic Theory of Finite Dimensional Normed Spaces. With an Appendix by M. Gromov. VIII, 156 pages. 1986.

Vol. 1201: Curvature and Topology of Riemannian Manifolds. Proceedings, 1985. Edited by K. Shiohama, T. Sakai and T. Sunada. VII, 336 pages. 1986.

Vol. 1202: A. Dür, Möbius Functions, Incidence Algebras and Power Series Representations. XI, 134 pages. 1986.

Vol. 1203: Stochastic Processes and Their Applications. Proceedings, 1985. Edited by K. Itô and T. Hida. VI, 222 pages. 1986.

Vol. 1204: Séminaire de Probabilités XX, 1984/85. Proceedings. Edité par J. Azéma et M. Yor. V, 639 pages. 1986.

Vol. 1205: B.Z. Moroz, Analytic Arithmetic in Algebraic Number Fields. VII, 177 pages. 1986.

Vol. 1206: Probability and Analysis, Varenna (Como) 1985. Seminar. Edited by G. Letta and M. Pratelli. VIII, 280 pages. 1986.

Vol. 1207: P.H. Bérard, Spectral Geometry: Direct and Inverse Problems. With an Appendix by G. Besson. XIII, 272 pages. 1986.

Vol. 1208: S. Kaijser, J.W. Pelletier, Interpolation Functors and Duality. IV, 167 pages. 1986.

Vol. 1209: Differential Geometry, Peñíscola 1985. Proceedings. Edited by A.M. Naveira, A. Ferrández and F. Mascaró. VIII, 306 pages. 1986.

Vol. 1210: Probability Measures on Groups VIII. Proceedings, 1985. Edited by H. Heyer. X, 386 pages. 1986.

Vol. 1211: M.B. Sevryuk, Reversible Systems. V, 319 pages. 1986.

Vol. 1212: Stochastic Spatial Processes. Proceedings, 1984. Edited by P. Tautu. VIII, 311 pages. 1986.

Vol. 1213: L.G. Lewis, Jr., J.P. May, M. Steinberger, Equivariant Stable Homotopy Theory. IX, 538 pages. 1986.

Vol. 1214: Global Analysis – Studies and Applications II. Edited by Yu.G. Borisovich and Yu.E. Gliklikh. V, 275 pages. 1986.

Vol. 1215: Lectures in Probability and Statistics. Edited by G. del Pino and R. Rebolledo. V, 491 pages. 1986.

Vol. 1216: J. Kogan, Bifurcation of Extremals in Optimal Control. VIII, 106 pages. 1986.

Vol. 1217: Transformation Groups. Proceedings, 1985. Edited by S. Jackowski and K. Pawalowski. X, 396 pages. 1986.

Vol. 1218: Schrödinger Operators, Aarhus 1985. Seminar. Edited by E. Balslev. V, 222 pages. 1986.

Vol. 1219: R. Weissauer, Stabile Modulformen und Eisensteinreihen. III, 147 Seiten. 1986.

Vol. 1220: Séminaire d'Algèbre Paul Dubreil et Marie-Paule Malliavin. Proceedings, 1985. Edité par M.-P. Malliavin. IV, 200 pages. 1986.

Vol. 1221: Probability and Banach Spaces. Proceedings, 1985. Edited by J. Bastero and M. San Miguel. XI, 222 pages. 1986.

Vol. 1222: A. Katok, J.-M. Strelcyn, with the collaboration of F. Ledrappier and F. Przytycki, Invariant Manifolds, Entropy and Billiards; Smooth Maps with Singularities. VIII, 283 pages. 1986.

Vol. 1223: Differential Equations in Banach Spaces. Proceedings, 1985. Edited by A. Favini and E. Obrecht. VIII, 299 pages. 1986.

Vol. 1224: Nonlinear Diffusion Problems, Montecatini Terme 1985. Seminar. Edited by A. Fasano and M. Primicerio. VIII, 188 pages. 1986.

Vol. 1225: Inverse Problems, Montecatini Terme 1986. Seminar. Edited by G. Talenti. VIII, 204 pages. 1986.

Vol. 1226: A. Buium, Differential Function Fields and Moduli of Algebraic Varieties. IX, 146 pages. 1986.

Vol. 1227: H. Helson, The Spectral Theorem. VI, 104 pages. 1986.

Vol. 1228: Multigrid Methods II. Proceedings, 1985. Edited by W. Hackbusch and U. Trottenberg. VI, 336 pages. 1986.

Vol. 1229: O. Bratteli, Derivations, Dissipations and Group Actions on C^*-algebras. IV, 277 pages. 1986.

Vol. 1230: Numerical Analysis. Proceedings, 1984. Edited by J.-P. Hennart. X, 234 pages. 1986.

Vol. 1231: E.-U. Gekeler, Drinfeld Modular Curves. XIV, 107 pages. 1986.

continued on page 187

Lecture Notes in Mathematics

Edited by A. Dold and B. Eckmann

1000

Heinz Hopf

Differential Geometry in the Large

Seminar Lectures New York University 1946
and Stanford University 1956

With a Preface by S.S. Chern

Second Edition

Springer-Verlag

Berlin Heidelberg New York London Paris Tokyo Hong Kong

Author

Heinz Hopf
Nov. 19, 1894 – June 3, 1971
Professor, Eidgenössische Technische Hochschule Zürich
1931 – 1965

AMS Subject Classifications (1980): 51 M 20, 52 A 25, 52 A 40, 53 A 05, 53 A 10, 53 C 21, 53 C 22, 53 C 45, 57 M 20

ISBN 978-3-540-51497-8 Springer-Verlag Berlin Heidelberg New York
ISBN 978-0-387-51497-0 Springer-Verlag New York Berlin Heidelberg

ISBN 978-3-540-12004-9 1. Auflage Springer-Verlag Berlin Heidelberg New York
ISBN 978-0-387-12004-1 1st Edition Springer-Verlag New York Heidelberg Berlin

2146/3140-543210 – Printed on acid-free paper

The editors are happy to make the famous 1946 and 1956 seminar lectures of Heinz Hopf on Geometry and Differential Geometry in the large available to the mathematical community. They are pleased to have this fine volume carry the number 1000 of the Lecture Notes in Mathematics series. They express their sincere thanks to all those who have contributed to the project: To Peter Lax and John Gray who wrote the original class notes; to the Mathematics Institutes of N.Y.U. and of Stanford University for the permission to rewrite and publish the notes; to S.S. Chern for suggesting the volume and writing a preface; to Konrad Voss and Karl Weber for carefully checking the old versions and correcting errors, partly using error lists made by Heinz Hopf himself; and to Rachel Boller for her excellent job in typing the final manuscript and drawing all illustrations.

Albrecht Dold
Beno Eckmann

PREFACE TO THE SECOND EDITION

The text of the Hopf Lecture Notes remains nearly unchanged. A number of misprints has been corrected, for which considerable help was given by WU TA-JEN of Nankai University at Tianjin, China, who also contributed a great number of valuable remarks.

One of the main questions discussed in Part-Two of the Hopf Lectures is the problem of finding all closed surfaces in E^3 with constant mean curvature (c.m.c.), the solution being given in these Lecture Notes for the genus 0 case and for the case of all simple closed surfaces of arbitrary genus (in which cases the round spheres are the only solutions), while "the question whether there exist closed surfaces of genus ≥ 1 with H=C and with self intersections ... remains unanswered" (p. 131). An exciting development began in 1986 with H.C. WENTE's proof of the existence of c.m.c. tori; this proof starts exactly at the point, where Heinz Hopf left the problem in 1950. In the meantime, not only have the c.m.c. tori been classified, but N. KAPOULEAS (1987) has also proved the existence of c.m.c. surfaces of arbitrary genus ≥ 3. The case of genus 2 still seems to make siffi- culties. For references see the paper of U. PINKALL and I. STERLING: On the classification of constant mean curvature tori, to appear in Annals of Mathematics (1989).

K. Voss
March 1989

PREFACE

These notes consist of two parts:

1) Selected Topics in Geometry, New York University 1946, Notes by Peter Lax.

2) Lectures on Differential Geometry in the Large, Stanford University 1956, Notes by J.W. Gray.

They are reproduced here with no essential change.

Heinz Hopf was a mathematician who recognized important mathematical ideas and new mathematical phenomena through special cases. In the simplest background the central idea or the difficulty of a problem usually becomes crystal clear. Doing geometry in this fashion is a joy. Hopf's great insight allows this approach to lead to serious mathematics, for most of the topics in these notes have become the starting-points of important further developments. I will try to mention a few.

It is clear from these notes that Hopf laid the emphasis on polyhedral differential geometry. Most of the results in smooth differential geometry have polyhedral counterparts, whose understanding is both important and challenging. Among recent works I wish to mention those of Robert Connelly on rigidity, which is very much in the spirit of these notes (cf. R. Connelly, Conjectures and open questions in rigidity, Proceedings of International Congress of Mathematicians, Helsinki 1978, vol. 1, 407-414).

A theory of area and volume of rectilinear polyhedra based on decompositions originated with Bolyai and Gauss. Gauss realized the delicacy of the problem for volumes, and Hilbert proposed in his famous "Mathematical Problems" that of "constructing two tetrahedra of equal bases and equal altitudes which can in no way be split into congruent tetrahedra..." (Problem no. 3). This was immediately solved by Max Dehn whose results, with some modifications, are presented in Part 1, Chapter IV of these notes. This work has been further pursued and treated by algebraic methods. For the modern developments I refer to C.H. Sah, Hilbert's third problem: Scissors congruence (Research Notes in Mathematics 33, Pitman, San Francisco 1979).

The main content of Part 2 consists of the study of Weingarten surfaces in the three-dimensional Euclidean space, particularly those for which the mean curvature or the Gaussian curvature is a constant. Important progress was recently made by Wu-Yi Hsiang, as he constructed many examples of hypersurfaces of constant mean curvature in the Euclidean space which are not hyperspheres; cf. Wu-Yi Hsiang, Generalized rotation hypersurfaces of constant mean curvature in the Euclidean spaces I (J. Differential Geometry 17 (1982), 337-356), and his other papers. But the simplest question as to whether there exists an immersed torus in the three-dimensional Euclidean space with constant mean curvature remains unanswered (the "soap bubble" problem).

Hopf's mathematical exposition is a model of precision and clarity. His style is recognizable in these notes.

S.S. Chern

March 1983

TABLE OF CONTENTS

Part One

Part Two

PART ONE

Selected Topics in Geometry

New York University 1946

Notes by Peter Lax

CHAPTER I

The Euler Characteristic and Related Topics

Section 1. The first topic to be discussed will be Euler's famous relation between the number of faces, edges and vertices of a convex polyhedron.

Definition. A convex 2-cell is a convex point set whose boundary consists of a finite collection of straight line segments (edges) which meet at points (vertices). A convex 3-cell is a convex point set whose boundary consists of a finite collection of convex two-cells.

Number of vertices of a three-cell will be denoted by e ;
Number of edges of a three-cell will be denoted by k ;
Number of two-cells of a three-cell will be denoted by f .

Euler's theorem states that for convex three-cells the following relation holds.

$$e - k + f = 2 \ . \tag{1.1}$$

A number of proofs of this theorem will be presented.

Section 2. First Proof (Legendre)

We are given P , a convex polyhedron; project its surface from an interior point into the surface of the unit sphere around that point. This may be done from theorems in the general theory of convex sets. This way we obtain a network on the surface of a sphere consisting of convex spherical polygons.

A Theorem on Spherical Polygons. The sum of the angles of a convex spherical polygon on the surface of a unit sphere is equal to $(n-2)\pi + A$, where n is the number of sides of the polygon and A its area.

This theorem can be proved by induction: for $n = 3$ it reduces to a well-known theorem in spherical trigonometry. To proceed from n to $n+1$ we subdivide the polygon into a triangle and a polygon of $(n-1)$ sides by means of a diagonal, which lies completely inside the original polygon because of its convexity.

The theorem holds for non-convex polygons as well but we shall not bother to prove it.

We return to our network consisting of convex spherical polygons.

For each polygon we write down the equation

$\Sigma\alpha_i = n\pi - 2\pi + A$ where α_i is an angle of the polygon.

We sum over all polygons, P_j. Then

$\sum_{i,j} \alpha_{ij} = 2\pi e$, since each vertex contributes a total angle of 2π.

$\Sigma\pi n_j = 2\pi k$, since each edge bounds two polygons and will be counted twice in summation of edges of the polygon.

$\sum_{j=1}^{f} 2\pi = 2\pi f$, since j goes from 1 to f.

$\sum_j A_j = 4\pi$, since every point of the sphere is covered once and only once, and the area of a unit sphere $= 4\pi$. But

$$\sum_{ij} \alpha_{ij} = \sum_j n_j\pi - \sum_j 2\pi + \sum_j A_j$$

$$2\pi e = 2\pi k - 2\pi f + 4\pi$$

dividing by 2π we get $e - k + f = 2$.

Section 3. Corollaries of Euler's Theorem

Let f_n denote the number of two-cells bounding the polyhedron which have n sides; obviously

(3.1) $$f = \sum_{n=3}^{\infty} f_n .$$

Since each edge bounds two polygons:

(3.2) $$2k = \sum_{n=3}^{\infty} n f_n ,$$

the total number of edges of all the polygons.

Let e_m be the number of vertices of the polyhedron at which m edges meet, obviously

(3.3) $$e = \sum_{m=3}^{\infty} e_m .$$

Since each edge contains two vertices:

(3.4) $$2k = \sum_{m=3}^{\infty} m e_m ,$$

the total number of edges emitted from all vertices.

Multiplying both sides of equation (1.1) by two and first substituting (3.1) and (3.2), (3.3), then (3.1), (3.3) and (3.4) we obtain

(1.1') $\Sigma 2e_m + \Sigma 2f_n - 4 = \Sigma nf_n$

(1.1") $\Sigma 2e_m + \Sigma 2f_n - 4 = \Sigma me_m$

$$\Sigma 4e_m + \Sigma 4f_n - 8 = \Sigma nf_n + \Sigma me_m ,$$

or

$$0 = 8 + \sum_{m=3}^{\infty} (m-4)e_m + \sum_{n=3}^{\infty} (n-4)f_n .$$

Putting all negative terms on the left side:

$$e_3 + f_3 = 8 + \sum_{m=5}^{\infty} (m-4)e_m + \sum_{n=5}^{\infty} (n-4)f_n .$$

Since all terms on the right are non-negative, it follows from this last expression that

(3.5) $e_3 + f_3 \geq 8$.

In particular (3.5) implies that

a) every convex polyhedron possesses either triangular faces, or vertices with exactly three edges, or possibly both.

Multiply (1.1') by two and add it to (1.1"):

$$\Sigma 6e_m + \Sigma 6f_n - 12 = \Sigma 2nf_n + \Sigma me_m ,$$

or

$$-12 = \sum_{3}^{\infty} (2n-6)f_n + \sum_{3}^{\infty} (m-6)e_m .$$

Arranging this equation so that both sides will contain positive terms only:

$$3e_3 + 2e_4 + e_5 = 12 + \sum_{n=4}^{\infty} (2n-6)f_n + \sum_{m=7}^{\infty} (m-6)e_m .$$

Since all terms on the right side are non-negative

$$3e_3 + 2e_4 + e_5 \geq 12 .$$

Similarly we can derive the inequality

$$3f_3 + 2f_4 + f_5 \geq 12 .$$

These last two inequalities imply that:

b) Every convex polyhedron must contain three, four or five-edged vertices.

c) Every convex polyhedron must contain triangles, or quadrilaterals or pentagons as faces.

Regular polyhedra. A regular polyhedron has the property that all its faces have the same number of sides n, and all its vertices have the same number of edges m. Therefore

$$e = e_m \ , \quad f = f_n \ .$$

By the previously derived results either $m = 3$ and $3 \leqslant n \leqslant 5$ or $n = 3$, and $3 \leqslant m \leqslant 5$ must hold.
Furthermore, the three equations:

$$2k = me = nf \ , \quad \text{and} \quad e - k + f = 2$$

determine e, k and f completely in terms of m and n. We tabulate all possible combinations

n	m	e	f	k	
3	3	4	4	6	Tetrahedron
3	4	6	8	12	Octahedron
3	5	12	20	30	Icosahedron
4	3	8	6	12	Hexahedron (cube)
5	3	20	12	30	Dodecahedron

We have thus proved that the five common regular polyhedrons are unique.

Section 4. Second Proof of Euler's Theorem (Steiner)

Consider in the plane a two-cell C with N sides subdivided into two-cells C_i. Let e, k, f denote the total number of vertices, edges and two-cells in C. e', k' and f' denote the number of interior vertices, edges and two-cells (i.e. those edges and vertices not on the boundary of C).

For each two-cell we have the well-known formula for the sum of the angles of a two-cell C_i with n sides and n angles each of α_j radians.

(4.1) $$\sum_{j=1}^{n} \alpha_j - n\pi + 2\pi = 0 \ .$$

We sum over all two-cells C_i and, as previously, we have

$\sum_{ij} \alpha_{ji} = 2\pi e' + \sum_C \alpha$, each interior vertex yields 2π to which we add $\sum_C \alpha$.

$\sum_i n_i \pi = 2\pi k' + \pi N$, each interior edge is on two 2-cells to which we add N .

$\sum 2\pi = 2\pi f'$, since $f = f'$.

Hence summing we get

(4.2) $$2\pi e' - 2\pi k' + 2\pi f' + \sum_C \alpha - \pi N = 0 .$$

Since (4.1) holds for C too, $\sum_C \alpha - \pi N + 2\pi = 0$. Substituting this into (4.2) and dividing it by 2π , we obtain the equation

(4.3) $$e' - k' + f' = 1 .$$

But, $e = e' + N$, $k = k' + N$, $f = f'$; substituting these into (4.3) we obtain

(4.3') $$e - k + f = 1 .$$

Let us consider next a subdivision of a straight line segment into subsegments, e, k, e', k' being defined as before. Since k', the number of subsegments is always one greater than the number of interior vertices e',

(4.4) $$e' - k' = -1$$

holds.

Similarly

(4.4') $$e - k = 1 .$$

(4.3') and (4.4') may be considered as Euler's Theorem for 2 and 1 dimensional cases.

Section 5. General Notion of Polyhedron

A polyhedron is a finite collection of two-cells with the following properties: Two 2-cells must be in one of the following three relations to each other:

a) They have no points in common;
b) They have one vertex in common;
c) They have one edge in common.

The characteristic $\chi(P)$ of a polyhedron P is defined as follows:

$$\chi(P) = e - k + f .$$

Subdivision of a polyhedron: By the subdivision of a polyhedron we mean a division of its two-cells by networks of edges and vertices in such a manner that when two 2-cells have an edge in common, any new vertex formed by the network of one cell on that edge of the original two-cell must coincide with a new vertex found in such a manner in the other two-cell.

Theorem: Let P_1, P_2 be two polyhedra, P_2 being a subdivision of P_1.

Then

(5.1) $$\chi(P_1) = \chi(P_2) .$$

<u>Proof</u>: Consider an edge of P_1 on which new vertices are introduced; the contribution of this open edge to $\chi(P_2)$ is, in view of (4.4), -1; the contribution of the same edge originally to $\chi(P_1)$ was also -1.

Consider an open two-cell of P_1 which is subdivided into two-cells; the contribution of the new vertices, edges and two-cells to $\chi(P_2)$ is, in view of (4.3), $+1$. The contribution of that same two-cell originally to $\chi(P_1)$ is $+1$ also. Since these are the only possible changes made by a subdivision, we see that the characteristic remains invariant.

The fact that the characteristic is invariant under subdivision is one of the most important tools in the classification of surfaces.

<u>Section 6</u>. We have shown that the characteristic is invariant under subdivision; this fact will be the basis of our third proof of Euler's theorem.

Consider two convex three-cells and project both of them from an interior point onto the surface of a sphere. This way we obtain two networks P and Q on the surface of the sphere. Consider S, the "combination" of P and Q. The two-cells of S are the non-empty intersections of two-cells of P and Q; the edges of S are the subdivision of edges of P and Q by points of intersections of P and Q; the vertices of S are the vertices of P and Q plus the points of intersections of P and Q. Thus S is a <u>subdivision of both P and Q</u>. By our previous result $\chi(S) = \chi(P)$ and also $\chi(S) = \chi(Q)$, hence

$$\chi(P) = \chi(Q) .$$

Hence all convex polyhedra have the same characteristic χ; specifically the characteristic of the tetrahedron is $4-6+4 = 2$; but then

$$\chi(P) = 2 \quad \text{or} \quad e - k + f = 2$$

must hold for <u>all</u> convex three-cells. This proves Euler's Theorem.

<u>Section 7</u>. <u>Surfaces of Higher Genus</u>

The previous considerations can be carried over to networks drawn on closed surfaces which are said to have different genus than the sphere. A surface of genus p can be obtained by cutting out $2p$ small circles from the surface of a sphere and connecting the punctures pairways so that they do not intersect. The surface shown in the diagram is of genus 2. A surface of genus 1 is called a torus. It may be

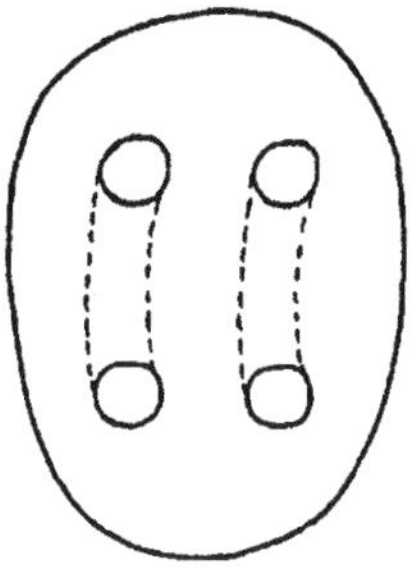

shown that all topologically equivalent closed orientable surfaces are of the same genus, and every closed orientable surface falls into a genus class.

<u>All networks</u> P <u>drawn on a surface of genus</u> p <u>have the same characteristic</u> $\chi(p)$.

This important theorem can be proven the same way we proved the special case $p = 0$ (sphere). Given networks P and Q , we obtain S by "combining" P and Q . Thus S is a subdivision of both P and Q , hence

$$\chi(S) = \chi(P) = \chi(Q) .$$

Thus we have shown that a characteristic χ can be associated with each surface of genus p . We wish to determine χ as a function of p . This we do inductively.

First we determine the characteristic of a torus. Two cuts in the manner indicated on the diagram will separate the torus into two parts;

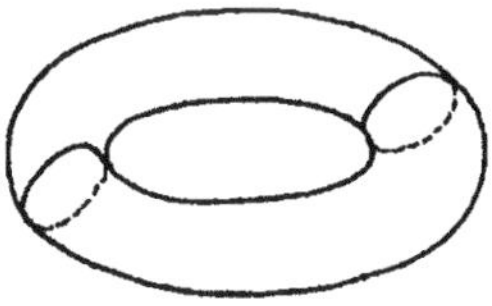

each of the parts will be equivalent to a sphere with two two-cells removed, therefore each part will have the characteristic zero. Putting the two parts together will not alter the characteristic since there were the same number of edges and vertices along the cut which will cancel themselves out. Thus we have shown that $\chi(\text{torus}) = 0$.

In case of a surface of genus p we cut it into two parts: One

will be a torus, minus one two-cell, the other a surface of genus (p-1), minus one two-cell. The characteristic of the former is -1, and that of the latter $\chi(p-1)-1$. As before putting the two parts together will not alter the characteristic, hence

$$\chi(p) = \chi(p-1) - 2 \quad . \tag{7.1}$$

Since

$$\chi(0) = \chi(\text{sphere}) = 2 \; , \tag{7.2}$$

it follows by induction from (7.1) and (7.2) that

$$\chi(p) = 2 - 2p \; . \tag{7.3}$$

Section 8. Application to the theory of Riemann surfaces

A surface Σ of genus p is represented by a single-valued continuous function on the sphere S.

All but a finite number of points (branch points) of Σ have a small neighborhood mapped in a one-to-one manner into a small region on S.

Furthermore we assume that each point which is not an image of a branch point is covered by the same number of sheets, let's say s times.

Let the number of branch points of order m be w_m, and put $W = \sum_m m w_m$.

The problem is to express p in terms of s and W.

Draw a net on S, stipulating that all branchpoints of the Riemann surface should be vertices of the net. Project the net on the Riemann surface.

Let e,k,f and ϵ,κ,φ be the number of vertices, edges and faces on S and on the Riemann surface respectively.

$$\varphi = sf \; , \quad \kappa = sk \; , \quad \epsilon = es - W$$

$$p = 1 - \chi(p)/2 = 1 - (\varphi-\kappa+\epsilon)/2 = 1 - s(f-k+e)/2 + W/2$$

$$= 1 - s + (1/2)W \tag{8.1}$$

which is the desired result.

One interesting consequence of (8.1) is that W is always even.

Consider the Riemann surface which the function $\sqrt{(\zeta-a_1)(\zeta-a_2)\ldots(\zeta-a_{2n})}$ gives rise to; here $W = 2n$ and $s = 2$. Hence the Riemann surface can be represented on a surface of genus $1-2+n = n-1$.

Problem 1. If instead of a sphere we relate the point transformation between a surface of genus p and one of genus q, how would formula (8.1) change?

Section 9. Role of the Euler Characteristic in the Theory of Vector Fields

Definition of Plane Vector Fields: To each point with a finite number of exceptions a direction is given; this direction is a continuous function of the plane except at the exceptional points; these points are called singular points.

Index of a Singular Point: Draw a circle around the singular point which is so small that it contains no other singularities of the vector field in its interior or on its circumference. Take an arbitrary point on the

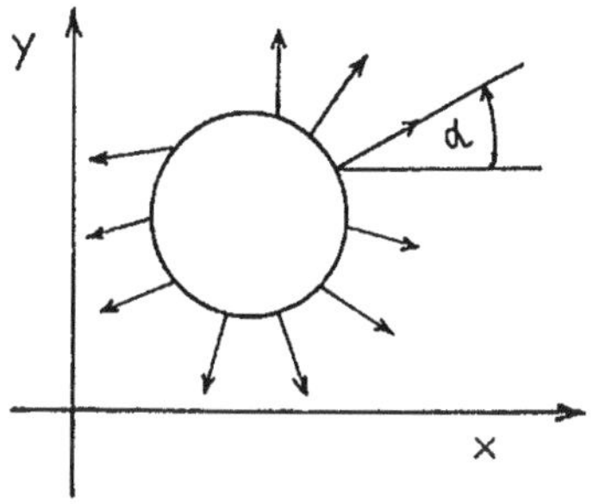

circumference of the circle and let α be the angle the vector at that point makes with the direction of the positive x axis. α is uniquely determined mod 2π. After fixing the value of α, the angle of a vector at any subsequent point on the circumference with the direction of x axis can be determined uniquely if we require that it be a continuous function of the arc length on the circumference of the circle. We consider the changing direction at each point and finally the total change of the α until we reach the original point. This will be a multiple of 2π.

$$\alpha(2\pi) - \alpha(0) = 2\pi j ,$$

where j is the integral multiple called the index.

a) The value of j is finite and it does not depend upon our initial point on the circumference of the circle.

b) The value of j does not depend upon the particular circle we choose as long as the region is free of other singular points. (Since two circles can be deformed into each other, and j changes continuously, hence remains a constant under such a deformation.)

Examples of singular points in vector fields:

(a) 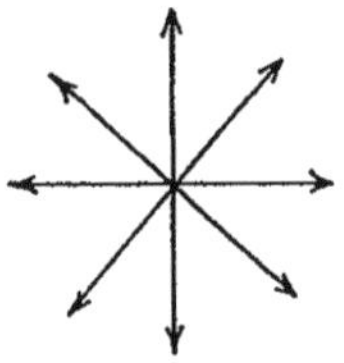 (b) 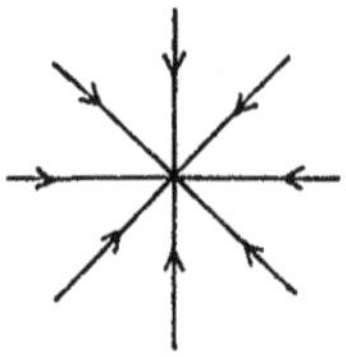 (c) 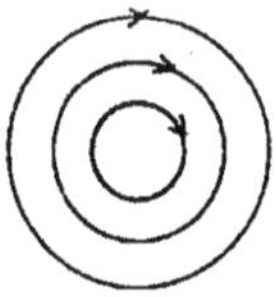

(d) 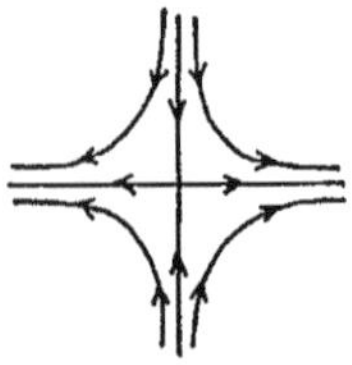 (e) 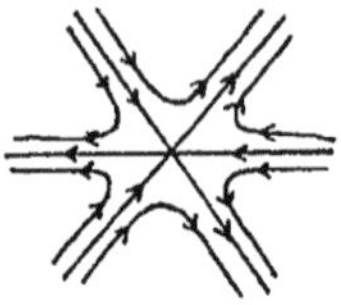 (f) 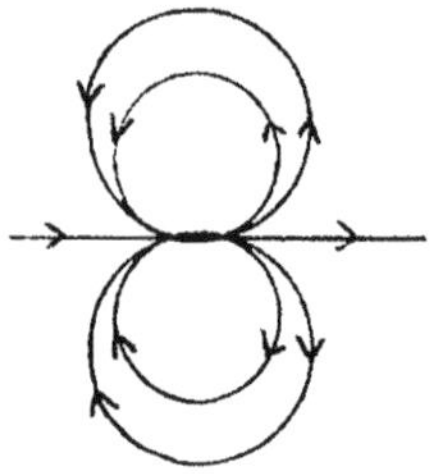

If we regard these fields as gradient fields,

(a) a source corresponds to a maximum,
(b) a sink, to a minimum,
(c) a center,
(d) a simple saddle-point,
(e) a monkey-saddle,
(f) a dipole.

The indices of these configurations are as follows:

(a)	1	(c)	1	(e)	-2
(b)	1	(d)	-1	(f)	2

The index was defined as $1/2\pi$ times the change of direction of the vector field with respect to a horizontal parallel vector field in going around a singular point x in a positive sense.

We observe that this change of angle in going around a regular point is zero. From this it follows that in the definition of the index the horizontal parallel vectorfield can be replaced by an arbitrary vector field which has no singular points in a sufficiently small neighborhood of x.

This generalized definition of the index is useful in studying Vectorfields on Closed Surfaces.

Definition: Let Σ be a closed surface possessing continuous first derivatives at every point; then a tangent plane exists at every point whose normal will vary continuously on the surface. Consider a field of tangent vectors (of unit length) defined and continuous at all but a finite number of points of this surface. These exceptional points are called singular points. A field with such properties will be called a regular vectorfield.

Index of a Singular Point: Take a small region around a singular point which contains no other singularities of the vectorfield; let the boundary of this region be a simple closed curve. Since the surface has continuous first derivatives at every point in every sufficiently small region we can define a non-singular tangent vector field. The index is defined as $1/2\pi$ the change in direction of the original vectorfield with respect to the local non-singular vectorfield.

We observe that the difference of the indices of two vectorfields F and F' on Σ at a point x can be defined without any reference to a local non-singular vectorfield. Namely the difference of the indices of F and F' at x is $\frac{1}{2\pi}$ times the change in the angle between the directions of the vectorfield F and F' in going around x in a positive sense. (The contribution of the local non-singular vectorfield cancels out.)

This observation will be of great importance in proving the following theorem:

The sum of the indices of all singularities of a regular vectorfield is equal to the characteristic of the surface.

$$\sum_r j_r = \chi(\Sigma) \quad . \tag{9.1}$$

Proof: The proof consists of two parts; first we show that $\sum_r j_r$ has the same value for all regular vectorfields on Σ.

We consider two regular vectorfields F and F'. We then subdivide Σ by a network in such a manner that there will be no singularities of either of the two fields on the edges or vertices of the network. Furthermore we make the two-cells of the network small enough, so as to have at most one singularity of F or F' in any one of them. Since there are a finite number of singularities this may easily be done.

The difference of indices $j_F - j_{F'}$ of the singularity in each two-cell is $\frac{1}{2\pi}$ times the change in the difference of the directions of F and F' in going around the boundary of the two-cell in a positive sense. But if we sum $j_F - j_{F'}$, over all two-cells, we find that this total sum is zero since every edge bounds two two-cells and in the course of calculating $\Sigma(j_F - j_{F'})$ each will be traversed in two opposite senses. Thus its contribution to $\Sigma(j_F - j_{F'})$ is zero.

Hence $\Sigma j_F = \Sigma j_{F'}$, and the sum of indexes is independent of the vectorfield.

The second part of the proof of (9.1) will consist of constructing a special vectorfield whose Σj will be calculated.

As in the first part divide Σ by a network in such a manner that all two-cells of the network are triangles.

In each triangle we introduce 4 additional points. The centers of the three sides and a point in the interior. The point in the interior is connected with the vertices of the triangle and the centers of the sides; these connecting lines are given a direction which points toward the interior point. Each half of the edges of the triangle are directed away from the vertex, from which it emanates (see diagram).

The original triangle is now broken up into six triangles; it is easy to see that it is possible to define a continuous vectorfield in

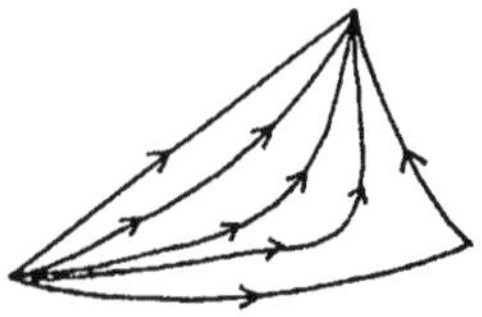

the interior of the triangles which will coincide on the boundary with the already prescribed directions.

Thus we have defined a vectorfield on Σ which has one singular point in each vertex, on each edge and in each two-cell of a network on Σ and all other points are regular points. Furthermore the index of the singularities in the vertices and in the two-cells are $+1$, since they are respectively sources and sinks while the index of the singularities on the edges is -1 , since they are of simple saddle-point type. If e,k and f denote the numbers of vertices, edges and two-cells of the network, we have

(9.1) $$\sum_r j_r = e + f - k = \chi(\Sigma)$$

Q.E.D.

With the aid of (9.1) we can answer the following important

question: On which surfaces exist vectorfields <u>free of singularities</u>?

If there are no singularities, $\Sigma j_r = 0$.
By (9.1) $\chi(\Sigma) = 0$ must hold. Therefore the only possible surface with the above mentioned property is the torus. By a simple construction (see diagram) we can show that such a vectorfield actually exists.

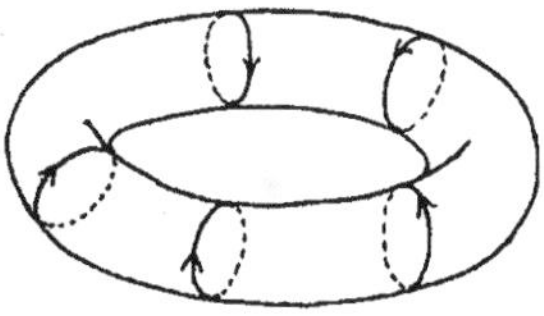

<u>Section 10</u>. In this lecture a purely combinatorial proof of Euler's theorem (due to Cauchy) will be presented.

Definition of <u>network</u>: A network in n-dimensional Euclidean space consists of a finite number of points (vertices) p and of straight line segments (edges) connecting some of these vertices, where no two edges have an interior point in common. An edge will be denoted by its endpoints $p_i p_j$.

A network on the surface of a sphere consists of a finite number of points (vertices) and of arcs of great circles (edges) connecting some of the vertices, where every pair of vertices is connected by at most one edge and no two edges have an interior point in common.

We introduce the following notation:

e = number of vertices
k = number of edges
f = number of open connected regions
c = number of components of the network

(A component is a connected part of the network which is not connected with the remainder of the network.)

Then the following equality holds:

(10.1) $$e - k + f = 1 + c .$$

In proving this theorem at some stage of the proof we must essentially use the hypothesis that the network is situated on the surface of a sphere (and not on a surface with genus $\neq 0$). We do this by using Jordan's theorem on polygonal arcs:

<u>Jordan's Theorem</u>: On the surface of a sphere every closed simple non-self intersecting polygonal arc divides the surface of the sphere

in exactly two open regions. Furthermore points on the opposite side of a boundary arc in a sufficiently small neighborhood of the arc belong to different regions.

The proof of (10.1) will be done by induction on the number of edges.

For $k = 0$ (10.1) holds because in that case the network consists of e isolated vertices; for this configuration

$$f = 1 \quad \text{and} \quad c = e \ ,$$

and we obtain

$$e - k + f = e - 0 + 1 = 1 + e = 1 + c \ .$$

We now show that if (10.1) holds for $k = n$ then it holds for $k = n+1$. This will be done by removing an edge from a network with $(n+1)$ edges. We will then show that in all cases the right and left side of (10.1) has decreased by the same amount. But first we must introduce the concept of a free vertex. A free vertex is one from which one edge emanates.

Lemma: If a network Q containing some edges does not contain free vertices then there exists at least one simple closed polygonal arc made up of edges of Q.

Proof of Lemma: Since Q contains edges, by hypothesis there exists a vertex p_1 in Q with an edge p_1p_2 emanating from p_1; since p_2 cannot be free, there will be an edge p_2p_3 different from p_2p_1 emanating from it. By the same process we get a chain of vertices $p_1, p_2, \ldots, p_n$ in which two consecutive vertices are connected by an edge and any three consecutive vertices are distinct. Since there are but a finite number of vertices in Q, we shall eventually come to a vertex p_n, for which $p_n = p_r$, $r < n$. We consider the first vertex that satisfies this condition: we then have a closed polygonal arc

$$p_r, p_{r+1}, p_{r+2} \cdots p_r$$

with at least 3 distinct edges in it. This proves the lemma.

Now we proceed with the induction.

Assume that (10.1) holds for all networks with $k = n$ edges and consider a network Q with $n + 1$ edges. We distinguish two cases:

(I) Q has a free vertex

(II) Q doesn't have free vertices

In case (I) let p be a free vertex, connected by an edge to another vertex $\dot{p}'$. If p' is also a free vertex, then the removal of the edge p , p' will

leave e unchanged,
leave f unchanged,
decrease k by one,
increase c by one.

We can verify that the right and left hand side of (10.1) changed by the same amount. - The case where p' is not free is trivial.

In case (II) we have no free vertices hence by our lemma there exists a closed polygon. Removing an edge of this polygon will

leave e unchanged,
decrease k by one,
leave c unchanged,
decrease f by one.

(Since points on opposite sides of the edge we have removed belonged to two different open regions.)

This completes the proof of (10.1).

Using (10.1) we shall derive two classical results of combinatorial topology.

Section 11. General notion of one-dimensional complex

A one dimensional complex is a finite collection of elements (vertices) $p_1, p_2, \ldots, p_n$ and pairs of elements (edges) $p_i p_j$ so that each vertex is situated on at least one edge.

Imbedding of a one dimensional complex: A one-dimensional complex C is said to be imbedded in an n-dimensional Euclidean space E_n (on S, the surface of a sphere) if there exists a network Q in E_n (on S) with the following properties: To every vertex p of C there corresponds one and only one vertex p' of Q ; to all edges $p_i p_j$ of C correspond simple disjoint polygonal arcs whose endpoints are p_i' and p_j' .

If E and K denote the number of vertices and edges of C , e and k the number of vertices and edges of Q , we notice that it follows from the definition that $e - k = E - K$.

In the following discussion we shall construct two complexes and prove that they cannot be imbedded on the surface of a sphere.

(A) Consider the complex A consisting of five vertices p_i ,

$i = 1,\ldots,5$ and 10 edges p_ip_j , $i = 1,\ldots,5$, $j = 2,\ldots,5$, $i<j$. Assume that it is possible to imbed it on the surface of a sphere; then we have a network Q for which (10.1) holds:

(11.1) $f = 2-(e-k)$ ($c = 1$, since Q is obviously connected)

$e - k = E - K = 5 - 10 = -5$; substituting this into (11.1) we obtain

$$f = 2 - (-5) = 7 .$$

Let i denote the total number of incidences of edges of the complex with the open regions of the network. Since every edge of A lies on some closed simple plygonal path, by Jordan's theorem each edge bounds two regions:*)

$$i = 2K .$$

On the other hand each open region has at least three edges of the complex on its boundary:

$$3f \leqslant i .$$

Substituting the values $f = 7$, $K = 10$ into

$$3f \leqslant i = 2K$$

we obtain

$$21 \leqslant i = 20$$

which is a contradiction. Hence the complex A cannot be imbedded on a sphere.

(B) Consider the complex B consisting of six vertices $p_1, p_2, p_3, q_1, q_2, q_3$ and nine edges p_iq_j , $i,j = 1,2,3$. In this case we have

$$E = 6 , K = 9 , f = 2 - (6-9) = 5 .$$

Let i denote the total number of incidences of edges of the complex with regions of the network. As before

$$i = 2K .$$

There are no triangular regions because if there were, at least two of the three vertices of the triangle would both be a p or both a q . However, no two p or q are connected by an edge while in a triangle all vertices are connected with each other. Hence each region has at least four edges on its boundary:

$$4f \leqslant i .$$

---

*) Actually all that we need here is that $i \leqslant 2K$.

Substituting the values $f = 5$, $K = 9$ into

$$4f \leq i = 2K$$

we obtain

$$20 \leq 18 .$$

Hence the complex B cannot be imbedded on the surface of a sphere since the assumption that it can lead to a contradiction.

We call a one-dimensional complex that cannot be imbedded in the surface of a sphere <u>singular</u>; a singular complex is called <u>irreducible</u> if eliminating any one of its edges would make it non-singular.

Kuratowski has shown that <u>the only two irreducible singular complexes are</u> A <u>and</u> B .

<u>Section 12</u>. The second result that we shall derive with the aid of (10.1) is a theorem of Cauchy that is of great importance in his proof of the rigidity of convex polyhedra.

Given a network Q on S (the surface of a sphere) not containing any free or isolated vertices. We divide its edges into two groups X and Y to form a network. The order of a vertex of Q is defined as the number of instances in which two neighboring edges emanating from the vertex belong to different groups. The order is always an even, non-negative number.

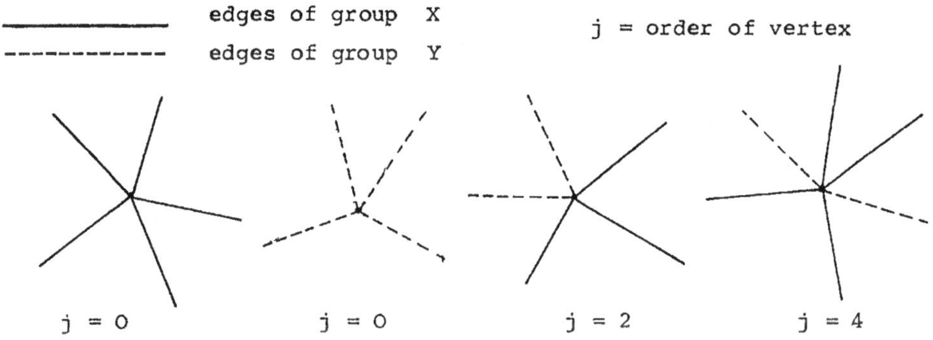

Call α the number of vertices of order 0 , β the number of vertices of order 2 and Γ_ν , $\nu = 1,2,...$ the number of vertices of order $(2+2\nu)$. Two adjacent emanating edges from the vertex form an angle. Let w be the total number of angles at all vertices, w_1 the number of angles whose two edges belong to the same group X or Y , and w_2 the angles whose edges are in different groups. Since there are no free vertices, each edge is on 4 angles and each angle has 2 edges

(12.1) $$2k = w = w_1 + w_2 \ .$$

Since at least two of the three edges of a triangle belong to the same group it follows that

$$w_1 \geqslant \text{number of triangles} = f_3 \ .$$

By formulas derived in section 3

$2k = \sum_{n=3}^{\infty} nf_n$, and $f = \sum_{n=3}^{\infty} f_n$; we see that

$$2k = 3f_3 + 4f_4 + 5f_5 + \ldots \geqslant 3f_3 + 4(f_4 + f_5 + f_6 + \ldots) =$$

$$= 3f_3 + 4(f - f_3) = 4f - f_3 \geqslant 4f - w_1 \ .$$

Adding this equality to (12.1) we obtain

$$2k \geqslant 4f - w_1$$
$$2k = w_1 + w_2$$

$$4k \geqslant 4f + w_2 \ , \text{ or}$$

(12.2) $$w_2 \leqslant 4(k-f) \ .$$

By (10.1) $k - f = e - 1 - c \leqslant e - 2$ and substituting this into (12.2) we obtain

(12.3) $$w_2 \leqslant 4e - 8 \ .$$

By definition

$$w_2 = 2\beta + \sum_{n=1}^{\infty} (2+2n)\Gamma_n \ ,$$

while the total number of vertices

$$e = \alpha + \beta + \sum_{n=1}^{\infty} \Gamma_n \ .$$

Substituting this into (12.3) we obtain

$$8 + 2\beta + \sum_{n=1}^{\infty} (2+2n)\Gamma_n \leqslant 4\alpha + 4\beta + 4 \sum_{n=1}^{\infty} \Gamma_n \ ,$$

or after rearrangement and division by 2

(12.4) $$2\alpha + \beta \geqslant 4 + \sum_{n=1}^{\infty} (n-1)\Gamma_n \geqslant 4 \ .$$

A vertex whose order is $\geqslant 4$ is called (for obvious reasons) a crosspoint. Inequality 12.4 means then that if the edges of a network Q on S are divided into two groups X and Y <u>there will always be at least two vertices which are not crosspoints</u>.

There is an analogy between this theorem and the theorem on the sum of indices of singularities of a vector field on S .

Section 13. Generalization of Euler's theorem to n dimensions

Before we attempt to establish and prove the generalization of Euler's theorem we would like to generalize the notions and lemmas that were used for proving it in 2 dimensions. In Legendre's and Steiner's proof we operated with theorems on the sums of angles of spherical and plane triangles and polygons. These theorems can be generalized to n dimensions; the results are elegant but are not as widely known as they should be.

A formula connecting the sum of the solid angles and dihedral angles of a tetrahedron was discovered by de Gua (1783).

Denote the solid angles at the vertices of the tetrahedron by α_i $(i = 1,2,\dots,4)$, the dihedral angles by β_i $(i = 1,2,\dots,6)$. Trihedral

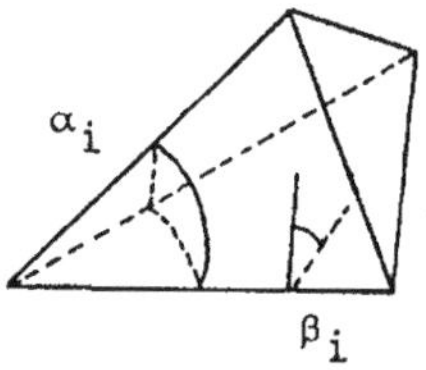

angles will be measured by the area cut off on a unit sphere with the vertex of the angle as origin. Dihedral angles will be measured in the usual way and are therefore equal to the spherical angles shown in the diagram. By the well known formula on the sum of angles of a spherical triangle (see section 2) we have for each vertex

$$\pi + \alpha_i = \Sigma\,\beta \tag{13.1}$$

where the summation is to be taken over all edges emanating from vertex i .

Summing (13.1) over all vertices $(i = 1,2,\dots,4)$ we obtain (since each dihedral angle contributes to <u>two</u> vertices)

$$\Sigma\,\alpha - 2\Sigma\,\beta = -4\pi\ .$$

Dividing this last equation by 4π we obtain

$$\frac{1}{4\pi}\,\Sigma\alpha - \frac{1}{2\pi}\,\Sigma\,\beta = -1\ . \tag{13.2}$$

This is de Gua's Formula.

If we change the unit of the solid angle so that the solid angle associated with the whole surface of a sphere is 1 , and the unit of the dihedral angle so that the angle between two planes at an angle of 2π is 1, (13.2) can be written in the more symmetrical form

$$\Sigma\alpha - \Sigma\beta = -1 \quad (13.2')$$

Furthermore, if we artificially associate with each face of the tetrahedron an angle γ_i $(i = 1,\ldots,4)$, $\gamma_i = 1/2$, (the solid angle of an internal hemisphere whose center is any point on the face), and with the interior another angle $\delta = 1$, (the solid angle of an internal sphere whose center is any interior point) (13.2') can be written in the still more symmetrical form

$$\Sigma\alpha - \Sigma\beta + \Sigma\gamma - \Sigma\delta = 0 \quad (13.2'')$$

It is this form that will be the easiest to generalize to n dimensions.

Section 14. Definition of an n dimensional simplex.

Given $n+1$ points in an n dimensional Euclidean space E_n where these $n+1$ points do not lie on any hyperplane of dimension lower than n, we introduce the following system of coordinates: One of the $(n+1)$ given points is chosen as the origin O; each point X of E_n can be written as a linear combination of the remaining n points $a_1, a_2, \ldots, a_n$:

$$X = \sum_{i=1}^{n} a_i x_i$$

where the x_i are real numbers. The set of points X whose coordinates satisfy the inequalities

$$\begin{cases} x_\nu \geq 0, & \nu = 1,2,\ldots,n \\ \sum_{\nu=1}^{n} x_\nu \leq 1 \end{cases} \quad (14.1)$$

is called an n-dimensional simplex with the vertices $O, a_1, a_2, \ldots, a_n$.

The boundary of the simplex is the set of points X for whose coordinates the inequalities (14.1) will be satisfied, with the sign of equality holding in at least one case. Consequently, the boundary consists of $n+1$ faces ($(n-1)$ dimensional simplices), each of these faces is contained in an $(n-1)$ dimensional hyperplane.

We pass $n+1$ hyperplanes parallel to the $(n+1)$ hyperplanes containing the faces of the simplex through a point of the space which may for sake of convenience coincide with the vertex O.

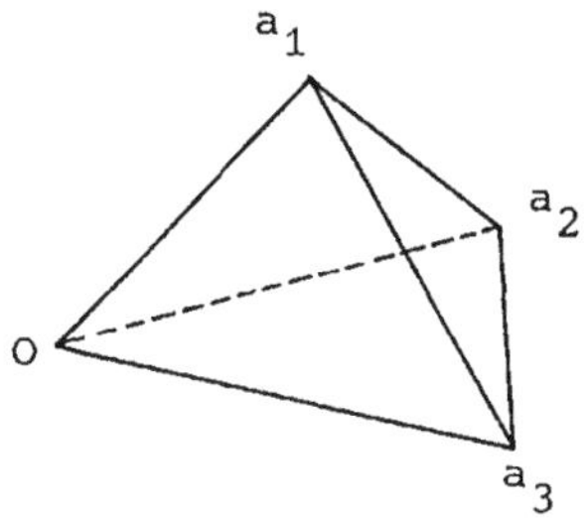

Each of these hyperplanes P_i , $i = 1,2,\ldots,n+1$ divides the whole space into two parts. The half space which, after a parallel translation of the plane into a position where it will contain one of the faces, contains the simplex is called the positive halfspace, the other the negative one.

Lemma I: No point of the space is on the positive (or negative) side of all hyperplanes P_i through O .

Proof of Lemma I: The positive halfspaces of each of the planes containing the n faces of the simplex passing through the vertex O are defined by the inequalities

$$(14.2) \qquad x_\nu > 0 , \qquad \nu = 1,2,\ldots,n .$$

The positive halfspace of the plane parallel to the plane containing the remaining face is defined by the inequality

$$(14.2') \qquad \sum_{\nu=1}^{n} x_\nu < 0 .$$

Obviously (14.2) and (14.2') cannot be satisfied simultaneously.

Q.E.D.

The intersection of (n-r) faces of the simplex is an r-cell of the simplex. We define the angle associated with this r-cell as the solid angle of that part of the (n-1) dimensional unit sphere about the origin which is on the positive side of the (n-r) planes parallel to the faces whose intersection defines the r-cell. (The unit of solid angle is chosen so that the solid angle of the full sphere is 1).

We introduce the following functions:

$$f_i(X) = \begin{matrix} 1 & \text{if } X \text{ is on the positive side of } P_i , \\ 0 & \text{otherwise.} \end{matrix}$$

S denotes the total surface area of the (n-1)-dimensional unit sphere.

Let α_r be the angle associated with the intersection of

$P_{i_1}, P_{i_2}, \ldots, P_{i_{n-r}}$.

We defined α_r as

$$\alpha_r = \frac{1}{S}\int f_{i_1}(X)\, f_{i_2}(X) \ldots f_{i_{n-r}}(X)\, dS \tag{14.3}$$

where the integration is extended over the unit sphere.

Let σ_r be the sum of the angles on all r-cells, i.e. $\sigma_r = \Sigma\,\alpha_r$. Lemma I of this section formulated in terms of the functions $f_i(X)$ states that the products

$$\prod_{i=1}^{n+1} f_i(X) \quad , \quad \prod_{i=1}^{n+1} (1-f_i(X))$$

are zero for all values of X .

Expanding the second one of these products and using the fact that the first one vanishes identically we obtain

$$1 - \Sigma f_i + \Sigma f_{i_1} f_{i_2} + \ldots + (-1)^n \Sigma f_{i_1} f_{i_2} \ldots f_{i_n} = 0 \ . \tag{14.4}$$

Integrating (14.4) term by term over the surface of the (n-1)-dimensional unit sphere and using (14.3) and the definition of σ_r we obtain

$$\sum_{r=0}^{n} (-1)^r \sigma_r = 0 \ . \tag{14.5}$$

This proof is due to Poincaré.

Since $\sigma_{n-1} = \frac{1}{2}(n+1)$ and $\sigma_n = 1$, due to their degeneracy, we have the corollary $\sum_{r=0}^{n-2} (-1)^r \sigma_r = (-1)^n (\frac{n-1}{2})$.

Section 15. Definition of n-dimensional convex polyhedron

An n-dimensional convex polyhedron is a convex point set whose boundary consists of a finite number of (n-1) dimensional convex polyhedra where any two (n-1) dimensional convex polyhedra have either no point or an r-cell in common.

Given an n-dimensional convex polyhedron we take an r-cell $(r \leqslant n)$ of it and about a point of this r-cell we construct an n-dimensional sphere. The n-dimensional solid angle of that part of the sphere which is inside the polyhedron is the angle associated with this r-cell.

Let σ_r denote the sum of the angles of all r-cells.

Let e_r , $r = 0,1,\ldots,n$, denote the number of r-cells of a convex polyhedron P. We define $\chi(P)$, the characteristic of P by

(15.1) $$\chi(P) = \sum_{r=0}^{n}(-1)^r e_r .$$

(This is a natural extension of the Euler characteristic in three dimensions.)

S^r denotes an r-dimensional sphere.

Let C^r be a convex r-cell whose interior is subdivided into r-cells. Let e_t denote the total number of t-dimensional elements, e'_t the number of interior t-dimensional elements. We define

(15.2) $$\chi(C^r) = \sum_{t=0}^{r}(-1)^t e_t , \quad \chi'(C^r) = \sum_{t=0}^{r}(-1)^t e'_t \, (r \geq t \geq 0) .$$

By the subdivision of an r-dimensional convex polyhedron we mean a subdivision of its $(r-1)$ cells in such a manner that if two $(r-1)$ cells have a t-cell in common, the subdivision of the two $(r-1)$ cells on this t-cell must coincide. Let P' stand for a subdivision of P.

After these preliminaries we state and shall prove the following four theorems:

(I) $\chi(S^r) = 1 + (-1)^r$, where $\chi(S^r)$ = the characteristic of <u>any</u> network on an r-dimensional sphere,

(II) $\chi(C^r) = 1$,

(III) $\chi'(C^r) = (-1)^r$,

(IV) $\chi(P) = \chi(P')$.

These four theorems will be proven by induction applied simultaneously to (I) and (III).

For $r = 1$ we can easily verify that (I) and (III) hold.

Assume that (I) and (III) hold for all $r < n$; we shall show that (I), (II), (III), (IV) follow for $r = n$.

Let α denote an angle associated with an r-cell of a convex n-cell, and $\gamma = \frac{1}{2} - \alpha$ the corresponding exterior angle; σ_r is the sum of the angles of its r-cells, and we let τ_r be the sum of the exterior angles of its r-cells. ($\tau_{n-1} = 0$).

We define
$$\Omega = \sum_{r=0}^{n}(-1)^r \sigma_r , \quad \Gamma = \sum_{r=0}^{n-2}(-1)^r \tau_r .$$

<u>Lemma II</u>. For every convex n-cell

(15.3) $$\Omega + \Gamma = \frac{1+(-1)^n}{2}$$

(15.4) $\Omega = 0$.

Proof of Lemma II. By summing the angles of all r-cells, on which by definition $\gamma = \frac{1}{2} - \alpha$ we see that

$$\tau_r = \frac{1}{2} e_r - \sigma_r, \quad \sigma_r = \frac{1}{2} e_r - \tau_r ;$$

$$\Omega = \sum_{r=0}^{n} (-1)^r \sigma_r = \sum_{r=0}^{n-1} (-1)^r \left(\frac{1}{2} e_r - \tau_r\right) + (-1)^n \text{ , since } \sigma_n = 1 ,$$

$$= 1/2 \sum_{r=0}^{n-1} (-1)^r e_r - \sum_{r=0}^{n-1} (-1)^r \tau_r + (-1)^n$$

and since the boundary of a convex n-cell can be considered as a network on an (n-1) sphere

$$\Omega = 1/2\, \chi(S^{n-1}) - \Gamma + (-1)^n$$

since we assumed that (I) holds for $r = n-1$,

$$\Omega + \Gamma = 1/2 \ (1+(-1)^{n-1}) + (-1)^n = 1/2(1+(-1)^n)$$

which proves (15.3).

We subdivide the (n-1)-cells of the convex n-cell into simplices (a possibility of such a subdivision is shown by the induction). The angle of any γ-cell introduced on an r-cell C^r $(n > r \geq \gamma)$ of the n-cell $= \alpha_r$, the angle of C^r ; since III is assumed to hold for $r < n$, the contribution of the γ-cells on C^r to Ω is $\chi'(C_r)\alpha_r = (-1)^r \alpha_r =$ contribution of C^r to Ω ; hence Ω remains invariant under a subdivision of the boundary.

We take a point interior to this subdivided convex n-cell and by connecting this point with all vertices on the subdivided boundary we subdivide the convex n-cell into n-dimensional simplices.

An interior r-cell of any of the simplices into which the original n-cell is subdivided consists of an (r-1)-cell of the boundary of the n-cell plus the interior point. Therefore if we denote by e_r the number of interior r-cells of the simplices, by e^*_r the number of r-cells on the boundary, the following equation holds:

(15.5) $e_0 = 1 \ , \ e_r = e^*_{r-1} \ , \ 1 \leq r \leq n$.

By (14.5), (15.4) holds for each of the simplices:

(15.6) $\sum_{r=0}^{n} (-1)^r \sigma_r = 0$.

We sum (15.6) over all n-dimensional simplices and since each interior r-cell contributes a total angle 1, we obtain

$$\Sigma\sigma_r = e_r + \sigma_r^* ,$$

where σ_r^* is the sum of angles of the r-cells of the subdivided n-cell; hence

$$\sum_{r=0}^{n} (-1)^r e_r + \sum_{r=0}^{n-1} (-1)^r \sigma_r^* = 0 . \tag{15.7}$$

Substituting (15.5) into (15.7) we obtain

$$1 - \sum_{r=0}^{n-1} (-1)^r e_r^* + \sum_{r=0}^{n-1} (-1)^r \sigma_r^* = 0 . \tag{15.8}$$

The boundary of a convex n-cell can be regarded as a network on an (n-1) sphere, and since (I) holds for $r = n-1$, the first sum in (15.8) can be replaced by $1 + (-1)^{n-1}$; hence we have by definition of Ω

$$\Omega = \sum_{r=0}^{n} (-1)^r \sigma_r^* = \sum_{r=0}^{n-1} (-1)^r \sigma_r^* + (-1)^n$$

which by (15.8)

$$= 1 + (-1)^{n-1} - 1 + (-1)^n = 0$$

which proves (15.4).

Using Lemma II we are able to prove (III) for $r = n$.

Let C^n be subdivided into convex n-cells C_i^n ; by Lemma II for each C_i^n (15.4) holds:

$$\sum_{r=0}^{n} (-1)^r \sigma_r^i = 0 . \tag{15.9}$$

Let e_r' denote the number of interior r-cells, σ_r^* the sum of the angles of the r-cells of C^n. Since each interior r-cell contributes a total angle 1 we obtain by summing (15.9) over all i :

$$\sum_{r=0}^{n} (-1)^r e_r' + \sum_{r=0}^{n-1} (-1)^r \sigma_r^* = 0 ,$$

hence

$$\chi'(C^n) = \sum_{r=0}^{n} (-1)^r e_r' = - \sum_{r=0}^{n-1} (-1)^r \sigma_r^* = (-1)^n - \Omega = (-1)^n$$

which proves (III) for $r = n$.

To prove (II) we observe that

$$\chi(C^n) = \chi'(C^n) + \chi(S^{n-1}) = (-1)^n + (1 + (-1)^{n-1}) = 1 .$$

To prove (IV) we observe that the contribution of an r-cell to $\chi(P)$ is $(-1)^r$, while the contribution of the same r-cell after subdivision to $\chi(P')$ is, by (III), also $(-1)^r$.

Since any two networks on S^n possess a common subdivision (see section 6), it follows that the characteristic of any network on S^n has the same value. Computing this value for the network defined by an (n+1) dimensional simplex, where

$$e_r = \binom{n+2}{r+1} \quad , \quad r = 0,\dots,n$$

$$\chi(S^n) = \sum_{r=0}^{n} (-1)^r \binom{n+2}{r+1} = 1 + (-1)^{n+2} - \sum_{r=0}^{n+2} (-1)^r \binom{n+2}{r}$$

$$= 1 + (-1)^n$$

which proves (I) and completes the induction.

Section 16. n-dimensional spherical simplices

Given an n-dimensional sphere of radius R, and (n+1) hyperplanes P_i which pass through its center. Each of these hyperplanes divides the whole space into two parts, one of which, arbitrarily chosen, will be designated as the positive, the other one as the negative half-space.

We define the functions

$$f_i(X) = \begin{cases} 1 & \text{if } X \text{ is on the positive side of } P_i, \\ 0 & \text{otherwise,} \end{cases} \quad i=1,2,\dots,n+1.$$

The closure of the set of points X on the surface of the sphere for which $f_1(X) = \dots = f_{n+1}(X) = 1$ is called an n-dimensional spherical simplex.

The set of points X on the surface of the sphere for which $f_1(X) = f_2(X) = \dots = f_{n+1}(X) = 0$ is another spherical simplex, antipodic to the first one and therefore congruent to it.

The intersection of (n-r) hyperplanes $P_{i_1}, P_{i_2}, \dots, P_{i_{n-r}}$ with the surface of the sphere defines an r-cell of the spherical simplex. An angle of this simplex α_r is defined as

$$(16.1) \qquad \alpha_r = \frac{1}{C_n R^n} \int f_{i_1}(X) f_{i_2}(X) \dots f_{i_{n-r}}(X)\, dS \, ,$$

where $C_n R^n$ is the surface area of the n dimensional sphere.

We consider the value of the product

$$(1-f_1(X)(1-f_2(X))\dots(1-f_{n+1}(X))$$

which is $= 1$ if X is in the antipodic simplex, and zero for all other values of X. If A_n denotes the area of the spherical simplex (= the area of the antipodic spherical simplex), we obtain

$$\int (1-f_1(X))(1-f_2(X))\dots(1-f_{n+1}(X))\,ds = A_n .$$

Evaluating the same integral term by term we obtain

(16.2) $\int (1-\Sigma f_i(X) + \Sigma f_{i_1}(X) f_{i_2}(X) \dots + (-1)^n \Sigma f_{i_1}(X) f_{i_2}(X) \dots f_{i_n}(X)$

$$+ (-1)^{n+1} f_1(X) f_2(X) \dots f_{n+1}(X))\,ds = A_n .$$

The value of the integral $\int f_1(X) \dots f_{n+1}(X)\,ds$ is $= A_n$, since the value of the product is 1 if X is in the spherical simplex, 0 otherwise. Using this fact and (16.1) we can rewrite (16.2) as

(16.3) $$C_n R^n \sum_{r=0}^{n} (-1)^r \sigma_r = (1+(-1)^n) A_n .$$

We distinguish two cases:

(i) n is even.

(ii) n is odd.

Using our previous symbols

$$\Omega = \sum_{r=0}^{n} (-1)^r \sigma_r , \quad \Gamma = \sum_{r=0}^{n-2} (-1)^r \tau_r$$

in case (i), (16.3) can be rewritten as

$$C_n R^n \Omega = 2A_n$$

or

(16.4) $$\Omega = 2/C_n \frac{A_n}{R^n} .$$

We can write (16.4) in terms of Γ : Substituting (15.3) into (16.4) we obtain

$$\frac{2}{C_n} \frac{A_n}{R^n} + \Gamma = 1 .$$

Replacing A_n by its definition as an integral we rewrite this last identity as

(16.5) $$\int \frac{ds}{R^n} + C_n/2\ \Gamma = C_n/2 .$$

(16.5) is the Gauss-Bonnet formula for n-dimensional spherical simplices.

(ii) For n odd (16.3) and (15.3) yield

(16.6) $$\Omega = \Gamma = 0 .$$

The forgoing derivation is due to Poincaré. He used (16.4) and (16.6) to obtain (15.4) by letting $R \to \infty$.

Using (16.4) and (16.6) we can easily generalize Legendre's proof of Euler's formula to n dimensions.

CHAPTER II

Selected Topics in Elementary Differential Geometry

Section 1. Curvature

Let $X(t) = (x(t), y(t), z(t))$ be a parametric representation of a curve in three-dimensional Euclidean space; assume that the functions $x(t), y(t), z(t)$ possess continuous second derivatives. The spherical image of $X(t)$ is constructed as follows: With any point $X(t_0)$ on the curve $X(t)$ we associate the point of intersection of the directed half-ray from the origin parallel to the directed tangent to $X(t)$ at $X(t_0)$ with the unit sphere about the origin. It follows from the differentiability properties of the curve $X(t)$ that its spherical image will possess continuous first derivatives.

We introduce s, the arc length of X from a fixed point s_0 as parameter. If $\Delta\sigma$ denotes the arc length of the spherical image between s and $s+\Delta s$, the limit $= \lim\limits_{\Delta s\to o} \frac{\Delta\sigma}{\Delta s} = \frac{d\sigma}{ds} = k$ exists and k is called the absolute curvature at s, or just curvature at s.

The total curvature between two points s_1 and s_2 is the length of the spherical image, $K = \int_{s_1}^{s_2} kds = \int_{\sigma_1}^{\sigma_2} d\sigma$.

The spherical image of a plane curve obviously lies on a great circle. We can give an orientation to this great circle by defining $\Delta\sigma$ as positive or negative according to the sense of rotation from s to $s+\Delta s$. Directed curvature is defined as $\lim\limits_{\Delta s\to o} \frac{\Delta\sigma}{\Delta s} = \frac{d\sigma}{ds} = k_1(s)$. Obviously $|k_1| = k =$ absolute curvature.

Again for the three dimensional case denote the angle which two tangent vectors at s_1 and s_2, respectively subtend by $u(s_1,s_2)$, $0 \leq u(s_1,s_2) \leq \pi$. This angle is equal to the spherical distance of the spherical images of the points $X(s_1)$, $X(s_2)$ on the unit sphere. By the geodesic property of arcs of great circles it follows that

$$u(s_1,s_2) \leq \int_{s_1}^{s_2} kds \ . \tag{1.1}$$

The sign of equality holds only in case of plane curves with a monotonically turning tangent if the total curvature of the arc is $\leq \pi$.

For plane curves we can define k_1 as above and the equation

$$u(s_1,s_2) = \int_{s_1}^{s_2} k_1 ds \tag{1.1'}$$

holds if the total curvature of the arc between s_1 and s_2 is $\leq \pi$.

Lemma I: If two plane curves $X_1(s)$ and $X_2(s)$ satisfy the following conditions

(a) They have the same curvature for all values of s,

(b) For one point s_o, $X_1(s_o) = X_2(s_o)$, $\dot{X}_1(s_o) = \dot{X}_2(s_o)$

then $X_1(s) = X_2(s)$ for all values of s, i.e. the two curves are identical.

Proof: Since the curvature is the derivative of the angle enclosed by the tangent and the x axis, it follows from (a) that $\tan^{-1}(\dot{y}_1/\dot{x}_1) = \tan^{-1}(\dot{y}_2/\dot{x}_2)$ for all values of s. This yields $\dot{y}_1/\dot{x}_1 = \dot{y}_2/\dot{x}_2$ and since $\dot{x}_1^2 + \dot{y}_1^2 = \dot{x}_2^2 + \dot{y}_2^2 = 1$, it follows that at any points s one of the equations

(1.2) $\dot{X}_1 = \dot{X}_2$, $\dot{X}_1 = -\dot{X}_2$

must hold. Since the vectors $\dot{X}_1$, $\dot{X}_2$ never vanish it follows that if one of the equations (1.2) holds for one value of s, it must hold for all values of s; but $\dot{X}_1(s_o) = \dot{X}_2(s_o)$ by (b), hence $\dot{X}_1(s) = \dot{X}_2(s)$ holds for all s. Integrating this equation from s_o to s and using from (b) the initial values $X_1(s_o) = X_2(s_o)$ we establish Lemma I.

We note that for curves in three dimensions the identity of the curvature does not imply the congruence of the curves:

Theorem A. (A. Schur) Let C and C' be two arcs of the same length with the endpoints a, b, a', b' respectively, $d = \overline{ab}$ and $d' = \overline{a'b'}$ denoting the distance of the endpoints; furthermore let $k(s)$ and $k'(s)$ denote the respective curvatures of C and C', where the parameter s is the arclength on C and C' measured from a and a' respectively.

If C is a plane curve and together with the chord connecting its endpoints forms a simple closed convex curve, and if at every point s, $0 \leq s \leq \ell$, $k'(s) \leq k(s)$ holds, then

$$d' \geq d ,$$

the sign of equality holding if and only if $C \cong C'$.

Proof: (E. Schmidt) Since C possesses a continuously turning tangent there exists a point s_1, $0 < s_1 < \ell$, where the direction of the tangent to C is parallel to the chord through a and b; let us call the point $C(s_1) = p$.

By hypothesis C together with the chord connecting a and b form a simple closed convex curve; therefore the angle enclosed by the tangent at $C(s)$ with the line through a and b is a monotonic function of s and its variation on the arcs $0 \leq s \leq s_1$, $s_1 \leq s \leq \ell$ is

$\leq \pi$. Therefore (1.1') is applicable:

$$u(s_1,s) = \left|\int_s^{s_1} k(s)\,ds\right| , \qquad 0 \leq s \leq \ell \quad .$$

We apply (1.1) to the curve C' :

$$u'(s_1,s) \leq \left|\int_s^{s_1} k'(s)\,ds\right|$$

and by the hypothesis $k'(s) \leq k(s)$ it follows that the last expression is

$$\leq \left|\int_s^{s_1} k(s)\,ds\right| = u(s_1,s)$$

Taking the cosine of both sides of this last inequality we obtain (since $\cos\theta$ is decreasing between 0 and π):

(1.3) $$\cos u'(s_1,s) \geq \cos u(s_1,s) \quad .$$

We integrate both sides of (1.3) from 0 to ℓ ; the integral of the right side is the length of the projection of the chord ab on the tangent at p ; since this tangent was chosen to be parallel to the chord, the length of the projection is equal to the length of the chord $= d$. The integral of the left side is the length d'' of the projection of the chord $a'b'$ on the tangent at p' ; therefore $d'' \leq$ the length of $a'b' = d'$; since the inequality (1.3) is preserved under integration, it follows that

(1.4) $$d' \geq d \quad .$$

The sign of equality holds if and only if:

(a) (1.1) as applied to C' the sign of equality holds, which implies that the two arcs $\widehat{a'p'}$ and $\widehat{p'b'}$ constituting C' are plane arcs.

(b) in $k'(s) \leq k(s)$ the sign of equality holds for all s, i.e. $k'(s) = k(s)$ for all s .

(c) $d'' = d'$.

By lemma I the equality of the curvature for plane curves implies the congruence of the curves. Therefore it follows from (a) and (b) that $\widehat{a'p'} \cong \widehat{ap}$, $\widehat{p'b'} \cong \widehat{pb}$. We want to show that $\widehat{a'p'}$ and $\widehat{p'b'}$ lie in the same plane.

Assume that $\widehat{a'p'}$ and $\widehat{p'b'}$ lie in two different planes; then their common tangent at p' must coincide with the line of intersection of these planes. But it follows from (c) that the tangent at p' must be parallel to the chord $a'b'$ which is possible only if both a' and b' lie on the line of intersection of the two planes. That means that the line through a', b' intersects the arc $\widehat{a'p'}$ at the point p' and since $\widehat{a'p'} \cong \widehat{ap}$, the line through the points a,b intersects C at p. This is contrary to the hypothesis that C together with the chord connecting its endpoints is a convex curve.

Thus we have shown that C' is a plane curve, consequently by (b) and Lemma I $C \cong C'$.

Q.e.d.

Section 2. Applications of Theorem A

Given two points $a'b'$ whose distance is $\overline{a'b'} = d'$. If we choose a value r so that $r \geq d/2$, it is possible to pass a circle of radius r through a' and b'. The two arcs connecting the points a' and b' both have a constant curvature $= 1/r$.

We consider now all curves C' connecting a' and b' whose curvatures $k(s)$ satisfy the inequality $k(s) \leq 1/r$. A theorem of H.A. Schwarz states that the arclength ℓ of such curves C' is either $\leq$ the length of the lesser arc or $\geq$ the length of the greater arc of the circle of radius r connecting a' and b'.

Proof: If $\ell \geq 2\pi r$, then ℓ is certainly $\geq$ the length of the greater arc; therefore we can restrict ourselves to the case $\ell < 2\pi r$. Let a,b be two points on a circle of radius r such that the length of one of the arcs $\widehat{ab} = \ell$; taking this arc as C in Theorem A, we obtain from (1.4) that

$$\overline{ab} \leq \overline{a'b'} = d' .$$

This means that if we draw the chord $a'b'$ parallel to ab, this chord will be nearer to the center of the circle than ab; this implies that

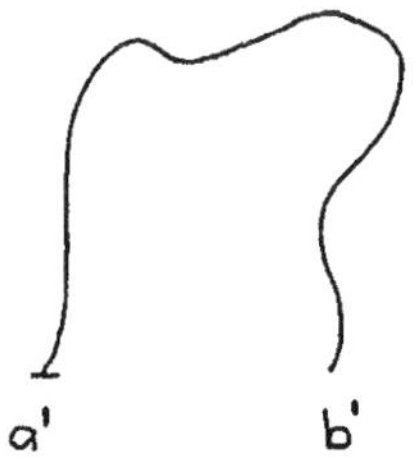

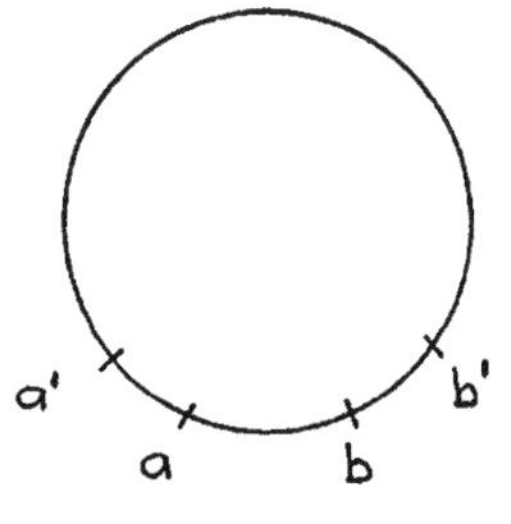

the length of the major arc $\widehat{a'b'}$ is $\leq$ the length of the major arc $\widehat{ab}$ and the length of the minor arc $\widehat{a'b'}$ is $\geq$ the length of the minor arc $\widehat{ab}$. The length of one of the arcs connecting a and b being $= \ell$ we see that the above statement is identical with Schwarz's theorem.

Q.e.d.

Section 3. Now we discuss the following problem: Of all arcs whose endpoints coincide and whose curvature $k(s)$ satisfies the inequality $k(s) \leq k_o$, which one has the smallest arc length? This question is motivated by the physical problem of finding the shortest piece of wire the endpoints of which can be brought together without breaking the wire, i.e. without increasing its curvature at any point beyond k_o .

We shall show that this curve is the circle of radius $\frac{1}{k_o}$. To prove this we assume that there exists a curve C' whose arc length ℓ is $< 2\pi/k_o$. Let a,b be two points on the circle of radius $1/k_o$ such that the length of one of the arcs $\widehat{ab} = \ell$. Taking this arc as C in Theorem A, we verify that the hypothesis of that theorem are satisfied, the circle being a convex arc and having a constant curvature $= k_o \geq k(s)$. Then we obtain that $d = \overline{ab} \leq$ distance of

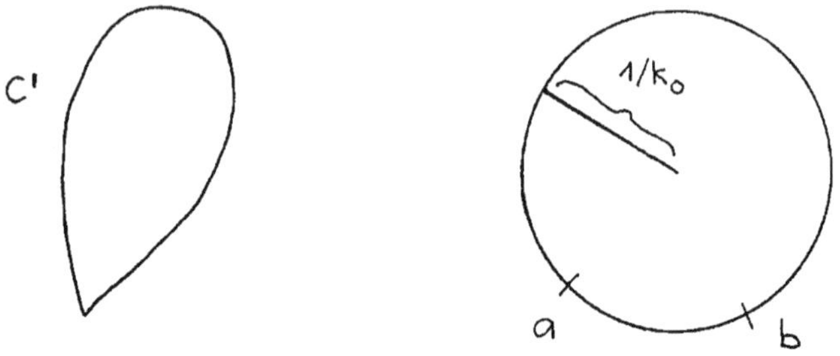

endpoints of $C' = 0$. But if $\ell < 2\pi/k_o$, the points a and b are distinct, i.e. $\overline{ab} > 0$; hence the assumption that there exists a C' whose arc length is $< 2\pi/k_o$ leads to a contradiction which proves the minimal property of the circle with radius $1/k_o$.

Section 4. Four vertex theorems

Take in the plane two closed simple convex curves C_1, C_2 which have the same arc length $L_1 = L_2 = L$. We make a one-to-one correspondence between points of C_1 and C_2 by taking one initial point arbitrarily on each of the two curves and introducing the arc lengths measured from those points as a parameter. The curvatures $k^1(s)$ and $k^2(s)$ are continuous and periodic functions of s with period L. Since the total curvature of closed simple convex curves is $= 2\pi$,

$$\int_0^L (k^1(s) - k^2(s))\,ds = \int_0^L k^1(s)\,ds - \int_0^L k^2(s)\,ds = 2\pi - 2\pi = 0 ,$$

from which it follows that unless $k^1(s) \equiv k^2(s)$, in which case $C_1 \cong C_2$, $k^1(s) - k^2(s)$ has to change sign at some point.

We shall prove that unless the two curves are congruent in which case there are no changes of sign <u>there are at least four changes of sign of</u> $k^1(s) - k^2(s)$ in the interval $0 \leq s \leq L$.

<u>Proof</u>: $k^1(s) - k^2(s)$ being a periodic function it follows that if the number of changes of signs is finite, it must be an even number; since the vanishing of $\int_0^\ell (k^1(s)-k^2(s))\,ds$ shows that the number of changes of signs is > 0, to prove our theorem all we have to show is that it is impossible to have exactly two changes of sign.

Assume that there are exactly two changes of sign occurring at $s = 0$ and $s = s_0$, i.e.

(4.1) $$k^1(s) \geq k^2(s) \quad \text{for} \quad 0 \leq s \leq s_0 ,$$

(4.2) $$k^1(s) \leq k^2(s) \quad \text{for} \quad s_0 \leq s \leq L .$$

If in Theorem A we identify $C_1(s)$, $0 \leq s \leq s_0$, with C, $C_2(s)$, $0 \leq s \leq s_0$ with C', from the conditions of simplicity and convexity imposed on C_1 and (4.1) we can verify that the hypotheses of Theorem A are satisfied. Hence from (1.4) we obtain

(4.3) $$d' \geq d ,$$

where

$$d' = \overline{C_2(s_0)C_2(0)} , \quad d = \overline{C_1(s_0)C_1(0)} .$$

On the other hand it follows from (4.2) that Theorem A is applicable with $C_2(s)$, $s_0 \leq s \leq L$ as C and $C_1(s)$, $s_0 \leq s \leq L$, as C'; for this case (1.4) yields an inequality which is the exact opposite of (4.3):

(4.3') $$d \geq d' .$$

(4.3) and (4.3') are at variance unless in both of them the sign of equality holds. But this is the case if and only if in both (4.1) and (4.2) the sign of equality holds, i.e. $k^1(s) = k^2(s)$ for all s which by Lemma I implies $C_1 \cong C_2$.

Q.e.d.

Given any closed convex curve C of arc length L, we draw a circle R with radius $L/2\pi$ and identify C and R with C_1 and C_2 of the previous theorem. The theorem states that $k^1(s) - k^2(s) = k(s) - 2\pi/L$ changes sign at least four times; since between two changes of sign a continuous function always has an extremal value, it follows

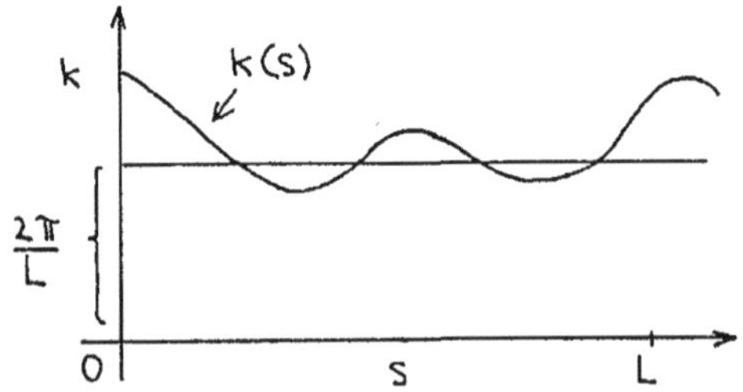

that the curvature of a closed simple convex curve has at least four extremals. This theorem is known as the <u>four vertex theorem</u> (Vierscheitel Satz).

<u>Section 5</u>. <u>Curves with discontinuously turning tangents</u>

Our discussion up to now applies only to curves which possess two continuous derivatives. The class of such curves will be referred to as class I . We shall now generalize some of our results to the class of curves which possess continuous second derivatives except at a finite number of points where a jump discontinuity in the first derivative may occur. The <u>exterior</u> angle formed by the right and left tangents at a point of discontinuity s will be denoted by $\alpha(s)$. This class of curves will be referred to as class II. A curve of class II is the sum of a finite number of curves of class I ; its spherical image is defined as the sum of the spherical images of the curves of class I that constitute it plus those minor arcs of great circle which connect the two different spherical images of the points where the first derivative has a discontinuity. The length of these connecting arcs is equal to the angles α enclosed by the right and left tangents; from this and the results of Section 1 of Chapter II, it follows that

(5.1) $$u(s_1,s_2) \leq \int_{s_1}^{s_2} d\sigma = \text{length of spherical image between } s_1 \text{ and } s_2 ,$$

the sign of equality holding for and only for plane curves with monotonically (though not necessarily continuously) turning tangents, if

$\int_{s_1}^{s_2} d\sigma \leqslant \pi$. The length of the spherical image can be written in the form:

$$(5.2) \qquad \int_{s_1}^{s_2} d\sigma = \int_{s_1}^{s_2} k(s)\,ds + \sum_{s>s_1}^{s<s_2} \alpha(s)$$

where $\Sigma\,\alpha(s)$ is the sum of all exterior angles enclosed by the right and left tangents at points of discontinuities between s_1 and s_2 . From (5.1) and (5.2) the following generalization of Theorem A follows:

Theorem A': If C and C' are curves having the same length, belonging to class II, and at every point of continuity $k(s) \geqslant k'(s)$, at points of discontinuity $\alpha(s) \geqslant \alpha'(s)$ holds, furthermore, if C together with the chord connecting its endpoints is a simple convex curve, then

$$(5.3) \qquad d' \geqslant d$$

where d and d' are the distances of the endpoints of C and C' respectively. The sign of equality holds if and only if $C \cong C'$.

We shall apply Theorem A' to prove an important lemma. We first introduce the following notation: Let G and G' be two corners in three-dimensional space with n faces each, F_ν and F'_ν respectively, $\nu = 1,2,\dots,n$; denote the face angles of F_ν and F'_ν by φ_ν and φ'_ν respectively. $\ell_{\nu+1}(\ell'_{\nu+1})$ denotes the edge of $G(G')$ bound by the two faces F_ν, $F_{\nu+1}(F'_\nu, F'_{\nu+1})$, $\nu = 1,2,\dots,n$ (where $F_{n+1} \equiv F_1$, $\ell_{n+1} \equiv \ell_1$); let the dihedral angle on ℓ_ν and ℓ'_ν be α_ν and α'_ν respectively.

O and O' denote the apexes of G and G' respectively.

Lemma II: If the following conditions are satisfied:

(a) $\varphi_\nu = \varphi'_\nu$ for $\nu = 1,2,\dots,n-1$

(b) $\alpha'_\nu \geqslant \alpha_\nu$ for $\nu = 2,3,\dots,n-1$

(c) G is convex, i.e. any straight line cuts the faces of the corner in at most two points.

then

$$\varphi'_n \geqslant \varphi_n ,$$

the sign of equality holding if and only if the sign of equality holds in (b) , $\nu = 2,3,\dots,n-1$.

Proof: We construct a plane P which cuts all edges of G but does not go through O ; let us denote the point of intersection of P with ℓ_ν by A_ν , $\nu = 1,2,\dots,n$. We construct the points A'_ν ,

$\nu = 1,2,\dots,n$ on the edges ℓ'_ν of G' so that $\overline{OA_\nu} = \overline{O'A'_\nu}$.

C and C' denote the polygons whose vertices are A_ν and A'_ν respectively, $\nu = 1,2,\dots,n$, and whose edges are the straight line segments connecting A_ν with $A_{\nu+1}$ and A'_ν with $A'_{\nu+1}$ respectively, $\nu = 1,2,\dots,n-1$.

It follows from (a) and the construction that the triangles $A_\nu O A_{\nu+1}$ and $A'_\nu O' A'_{\nu+1}$ are congruent for $\nu = 1,2,\dots,n-1$, hence $\overline{A_\nu A_{\nu+1}} = \overline{A'_\nu A'_{\nu+1}}$. By simple trigonometry it follows from (b) that the angles of C are $\geqslant$ the corresponding angles of C'. Furthermore, it follows from (c) that C, taken together with the chord connecting its endpoints is a simple convex curve. Hence C and C' satisfy all the hypotheses of Theorem A'; (5.3) yields

$$\overline{A'_n A'_1} \geqslant \overline{A_n A_1} \tag{5.4}$$

from which it follows by simple trigonometry that

$$\varphi'_n = \sphericalangle A'_n O' A'_1 \geqslant \sphericalangle A_n O A_1 = \varphi_n \;.$$

The sign of equality holds if and only if $\alpha_\nu = \alpha'_\nu$ for $\nu = 2,3,\dots,n-1$, i.e. $G \cong G'$.

Q.e.d.

Let G and G' be two <u>convex</u> corners with n faces each such that $\varphi_\nu = \varphi'_\nu$ for $\nu = 1,2,\dots,n$. We divide the edges of G into three classes

(a) The class of edges ℓ_κ for which $\alpha_\kappa > \alpha'_\kappa$
(b) The class of edges ℓ_λ for which $\alpha_\lambda < \alpha'_\lambda$
(c) The class of edges ℓ_μ for which $\alpha_\mu = \alpha'_\mu$.

We define the function $I(\nu)$:

$$I(\nu) = \begin{cases} 1 & \text{if } \ell_\nu \text{ belongs to (a)} \\ -1 & \text{if } \ell_\nu \text{ belongs to (b)} \\ 0 & \text{if } \ell_\nu \text{ belongs to (c)} \;. \end{cases}$$

The number of changes of sign of $I(\nu)$ as ν assumes its values $1,2,\dots,n-1, n, 1$ consecutively is called the <u>index</u> of (G,G') and is denoted by j. Obviously j is a non-negative even integer.

<u>Lemma III</u>: $j \geqslant 4$ unless $G \cong G'$.

<u>Proof</u>: All we have to show is that $j = 0$ and $j = 2$ implies $G \cong G'$.

(i) $j = 0$ means that either (a) or (b) is empty. If (a) is empty, Lemma II implies $\varphi'_n \geqslant \varphi_n$, where the sign of equality holds if and only if (b) is empty also. But $\varphi_n' = \varphi_n$ by hypothesis, hence (b) is empty. Similarly upon the application of Lemma II the emptiness of (b) implies the emptiness of (a) ; but if all edges belong to (c), $G \cong G'$.

(ii) $j = 2$ means that

(5.5) $\alpha_\nu' \geqslant \alpha_\nu$ holds for $\nu = 1,2,\ldots,m$

and

(5.6) $\alpha_\nu' \leqslant \alpha_\nu$ holds for $\nu = m+1, m+2,\ldots,n$

where m is some integer between 1 and n .

By identifying the corner formed by the edges ℓ_ν , $\nu = 1,2,\ldots,m$ with G , and the corner formed by the edges ℓ_ν' , $\nu = 1,2,\ldots,m$ with G' , of Lemma II we obtain by that Lemma that

(5.7) $\sphericalangle \ell_1'\ell_m' \geqslant \sphericalangle \ell_1\ell_m$

where $\sphericalangle \ell_1'\ell_m'$ and $\sphericalangle \ell_1\ell_m$, denote the angles enclosed by ℓ_1' and ℓ_m' , and ℓ_1 and ℓ_m respectively.

By identifying the corner formed by ℓ_ν , $\nu = m+1, m+2,\ldots,n$ with G' , and the corner formed by the edges ℓ_ν' , $\nu = m+1, m+2,\ldots,n$ with G , of Lemma II we obtain by that lemma

(5.8) $\sphericalangle \ell_1\ell_m \geqslant \sphericalangle \ell_1'\ell_m'$.

(5.7) and (5.8) are at variance unless in both of them the sign of equality holds. But then the corners identified with G and G' in Lemma II are congruent, hence G and G' of Lemma III are congruent also. Q.e.d.

This last result is an essential lemma in Cauchy's famous theorem on Rigidity of convex polyhedra: Given P_1 and P_2 , two convex polyhedra whose faces are in a one-to-one correspondence, corresponding faces being congruent and joined in the same order. Then the corresponding dihedral angles are equal, i.e. P_1 and P_2 are congruent.

Proof: We consider two classes of edges of P_1 :

(a) Those where the dihedral angle is > the dihedral angle at the corresponding edge of P_2 .

(b) Those where the dihedral angle is < the dihedral angle at the corresponding edge of P_2 .

We project P_1 from an interior point into the surface of a sphere about the interior point; the projection of those edges which belong either to (a) or to (b) forms a network Q on the sphere. This network will be empty if (a) and (b) are empty.

According to Sec.12 of Chapter I if the edges of a non-empty network Q are divided into two classes there always exist at least two vertices whose order j is < 4. Since by Lemma III $j \geq 4$ for every vertex, Q is empty. Hence (a) and (b) are empty also, i.e. all corresponding dihedral angles are equal.

Q.e.d.

Section 6. We shall now present a shorter proof of the four vertex theorem, due to Herglotz.

First we shall state without proof an elementary lemma from the theory of functions of real variables.

Lemma IV. If a continuous function has no extremal values in the interval (a,b), then it is monotonic in this interval.

Now let the simple closed convex curve C be represented in some rectangular coordinate system:

$$(x(s),y(s)) , \quad 0 \leq s \leq L$$

where L is the total arc length.

Let $\theta(s)$ denote the angle of the tangent at s : then

$$\dot{x} = \cos\theta , \ \dot{y} = \sin\theta .$$

Differentiating the first equation and substituting the second expression in the formula we obtain

$$\ddot{x} = -\sin\theta \cdot \dot{\theta} = -\dot{y}k$$

since $\dot{\theta} = k$ by definition. This last expression shows that $\dot{y}k$ is the derivative of a continuous, periodic function, hence its integral over the period vanishes:

$$(6.1) \qquad \int_0^L \dot{y}k \, ds = [-\dot{x}]_0^L = 0$$

Now let us assume that $k(s)$ has, beside a maximum at $s = 0$ and a minimum at $s = s_0$, no other extremals. We choose our coordinate axes so that the x-axis coincides with the line through $(x(0), y(0))$ and $(x(s_0), y(s_0))$. C being convex this line doesn't intersect the curve at any other point, therefore for $0 < x < s_0$: $y(s) > 0$, and for $s_0 < s < L$: $y(s) < 0$; furthermore by Lemma IV $k(s)$ is monotonic non-decreasing from $s = 0$ to $s = s_0$, and monotonic non-increasing from

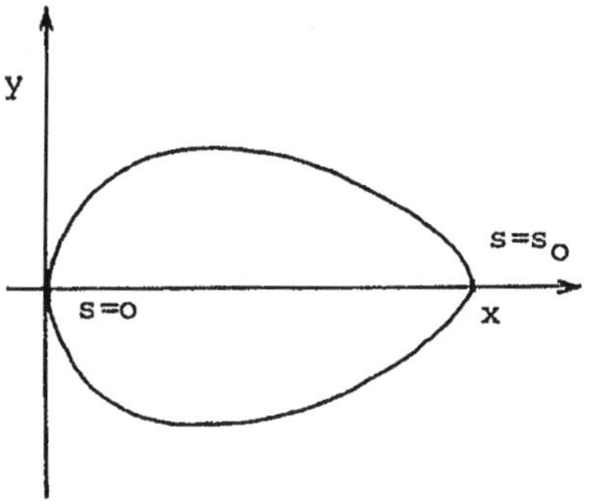

$s = s_o$ to $s = L$. By the second mean value theorem there exists ξ_1 , $0 < \xi_1 < s_o$ such that

$$\int_0^{s_o} k\dot{y}ds = k(0)\int_0^{\xi_1}\dot{y}ds + k(s_o)\int_{\xi_1}^{s_o}\dot{y}ds = k(0)y(\xi_1)$$

$$+ k(s_o)[y(s_o)-y(\xi_1)] = [k(0) - k(s_o)]y(\xi_1) .$$

Similarly there exists a ξ_2 , $s_o < \xi_2 < L$ such that

$$\int_{s_o}^{L} k\dot{y}ds = k(s_o)\int_{s_o}^{\xi_2}\dot{y}ds + k(L)\int_{\xi_2}^{L}\dot{y}ds = k(s_o)y(\xi_2) +$$

$$+ k(L)[y(L) - y(\xi_2)] = [k(s_o) - k(L)]y(\xi_2) = [k(s_o) - k(0)]y(\xi_2) .$$

Adding these two equations we obtain

(6.2) $$\int_0^L k\dot{y}ds = [k(0) - k(s_o)][y(\xi_1) - y(\xi_2)]$$

and by (6.1) this last expression is zero. But the first factor of the right side of (6.2) $\neq 0$ unless $\max k(s) = \min k(s)$ i.e. $k(s)$ is a constant. The second factor is always > 0 , since $y(\xi_1) > 0 > y(\xi_2)$ for the values of ξ_1 and ξ_2 that were chosen. Hence their product cannot be zero which shows that the assumption that $k(s)$ has only two extremals and $k(s) \not\equiv$ constant leads to a contradiction.

Q.e.d.

The four-vertex theorem holds for all simple closed curves but we shall not give a proof of this generalization.

Section 7. The total curvature of a simple closed convex curve is $= 2\pi$; we shall generalize this result.

For a plane curve C we have defined k_1 as the oriented curvature; we are going to show that for <u>simple</u> closed plane curves

$$(7.1) \qquad \int_0^L k_1 ds = \oint d\theta = 2\pi ,$$

if the curve is oriented so that the order of its interior is $+1$.

<u>Proof</u>: For $0 \leq s_1 < s_2 < L$ we define $V(s_1,s_2)$ as the argument of the vector pointing from $C(s_1)$ to $C(s_2)$; since C has no double points and it possesses a continuous tangent, $V(s_1,s_2)$ is continuous in the closure of its domain of definition. We choose $s = 0$ so that the horizontal supporting line touches C at $C(0)$. In view of the positive orientation of C the positive direction of the tangent at $C(0)$ is as indicated by the arrow on the accompanying diagram. The vector field $V(s_1,s_2)$ is defined in the triangle indicated on the diagram and is continuous there. Since the variation of the argument of a continuous vector field around a closed path is zero, it follows that $\int_{s=0}^{s=L} d\theta$ = the

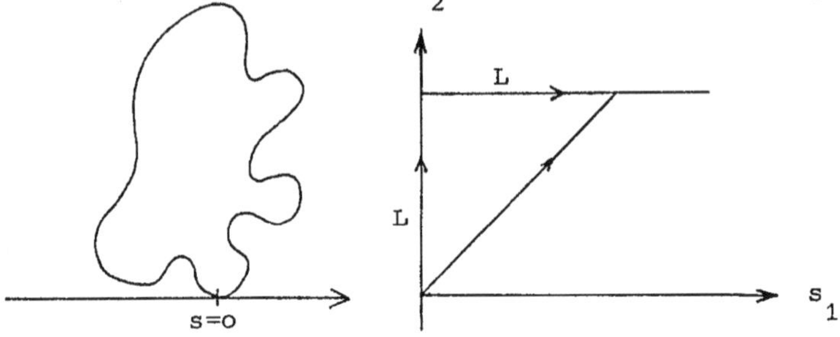

variation of $V(s_1,s_2)$ from $(0,0)$ to (L,L) along the hypothenuse is = the sum of the variation of $V(s_1,s_2)$ from $(0,0)$ to $(0,L)$ and $(0,L)$ to (L,L) along the legs of the triangle.

To evaluate the variation of $V(s_1,s_2)$ from $(0,0)$ to $(0,L)$ along the leg of the triangle we observe that, since the tangent at $C(0)$ is a supporting line of C_1 , $0 \leq V(0,s) \leq \pi$; since $V(0,s)$ is thus restricted to this sector, its variation as s goes from 0 to L is $= V(0,L) - V(0,0) = \pi$; similarly we find that the variation of $V(s_1,s_2)$ along the other leg is also $= \pi$. Adding these quantities we establish the validity of (7.1).

<u>Section 8</u>. It follows from (7.1) that for plane curves $\int_0^L k\, ds = \int_0^L |k_1| ds \geq \int_0^L k_1 ds = 2\pi$. We shall now demonstrate that the inequality

$$(8.1) \qquad \int_0^L k\, ds \geq 2\pi$$

holds for all closed curves, i.e., the total curvature of any closed curve is $\geq 2\pi$.

Given any curve belonging to class I of total length L , $(x(s), y(s), z(s))$, $0 \leq s \leq L$, we consider its spherical image $S(s) = (\dot{x}(s), \dot{y}(s), \dot{z}(s))$, $0 \leq s \leq L$; $S(s)$ is a closed rectifiable spherical curve. Furthermore

$$(8.2) \qquad \int_0^L S(s)\,ds = \int_0^L (\dot{x}(s), \dot{y}(s), \dot{z}(s))\,ds = 0$$

because x, y and z are periodic functions of s with period L .

<u>Lemma V</u>: Given a set of k vectors in m dimensional space $X_1, X_2, \ldots, X_k$, satisfying the following equation:

$$(8.3) \qquad \sum_{i=1}^{k} a_i X_i = C$$

where the a_i are <u>positive</u> numbers, then: It is possible to select n vectors $X_{i_1}, X_{i_2}, \ldots, X_{i_n}$ from the given k vectors and determine n <u>positive</u> numbers $b_1, b_2, \ldots, b_n$ such that

$$\sum_{j=1}^{n} b_j X_{i_j} = \sum_{j=1}^{k} a_j X_j \; , \quad \sum_{j=1}^{n} b_j = \sum_{j=1}^{k} a_j$$

and

$$(8.4) \qquad n \leq m+1 \; .$$

<u>Proof</u>: We shall prove Lemma V by induction on k . Lemma V holds for $k = 1$. Assume that Lemma V holds for all values of $k \leq k_0$; we shall show that then it holds for $k = k_0 + 1$ also.

If $k_0 \leq m$, then Lemma V is trivially satisfied by $b_j = a_j$, $j = 1, \ldots, k_0 + 1$, since $n = k_0 + 1$ satisfies (8.4).

If $k_0 > m$, we consider the system of $m+1$ homogeneous equations for $k_0 + 1$ unknowns $\lambda_1, \lambda_2, \ldots, \lambda_{k_0+1}$:

$$(8.5) \qquad \sum_{i=1}^{k_0+1} \lambda_i X_i = 0$$

$$\sum \lambda_i = 0 \; .$$

Since the number of unknowns, $k_0 + 1$, is greater than the number of equations $m+1$, the system (8.5) will always have a non-trivial set of solutions.

$$\lambda_1, \lambda_2, \ldots, \lambda_{k_0+1} \; .$$

Of these numbers some - let us say the first μ - will be positive, the others non-positive:

$$\lambda_i > 0 \quad \text{for} \quad i = 1,2,\dots,\mu \quad , \quad \lambda_i \leq 0 \quad \text{for} \quad i = \mu+1,\dots,k_0+1 \; .$$

We choose λ_ν , $1 \leq \nu \leq \mu$ so that

$$(8.6) \qquad 0 < t = \frac{a_\nu}{\lambda_\nu} \leq \frac{a_j}{\lambda_j} \quad \text{for} \quad j = 1,2,\dots,\mu \; .$$

We write

$$a_i' = a_i - \lambda_i t \; ;$$

then

$$a_\nu' = a_\nu - \lambda_\nu t = 0 \; , \; a_j' = a_j - \lambda_j t \geq 0 \quad \text{for} \quad j = 1,2,\dots\mu$$

because of (8.6) and

$$a_j' = a_j - \lambda_j t > 0 \quad \text{holds for} \quad j = \mu+1,\dots,k_0+1$$

since for these values of j , $\lambda_j \leq 0$.

Multiplying the first of equations (8.5) by t and subtracting it from (8.3) we see that

$$(8.7) \qquad \sum_{i=1}^{k_0+1} a_i' X_i = C$$

and from the second of equations (8.5) we see that

$$(8.8) \qquad \sum_{i=1}^{k_0+1} a_i = \sum_{i=1}^{k_0+1} a_i' \; .$$

Of the set of non-negative numbers $\{a_i'\}$ we select those a_j' which are positive; since $a_\nu' = 0$, the number of positive a_j'-s is $\leq k_0$. Since equations (8.7) and (8.8) remain unchanged if we omit those a_j which are zero, we have succeeded in selecting from the original k_0+1 vectors n vectors, $n \leq k_0$, such that a linear combination of these vectors with positive coefficients $= \sum_{i=1}^{k_0+1} a_i X_i$ and the sum of the coefficients in this linear combination $= \sum_{i=1}^{k_0+1} a_i$. Thus we have reduced the case $k = k_0+1$ to the case $k = n \leq k_0$, for which Lemma V holds by the assumption of the induction.

Q.e.d.

From (8.2) and the definition of Riemann integral it follows that to any given ε we can find N such that

$$(8.9) \qquad \left| \sum_{\nu=1}^{N} \frac{L}{N} \, s\left(\frac{\nu L}{N}\right) \right| < \varepsilon \; .$$

Then by the special case $m = 3$ of Lemma V we can find $n \leq 4$ points $s\left(\frac{\nu_i L}{N}\right)$, $i = 1,\dots,n$ and n positive numbers b_i , $i = 1,\dots,n$ such that

$$\sum_{\nu=1}^{N} \frac{L}{N} \, s\left(\frac{\nu L}{N}\right) = \sum_{1}^{n} b_i s\left(\frac{\nu_i L}{N}\right) ,$$

$$\sum_{\nu=1}^{N} \frac{L}{N} = L = \sum_{i=1}^{n} b_i \; .$$

We let ε assume a sequence of values tending to zero; we shall correspondingly have a sequence of values N_ε, $n(\varepsilon) \leq 4$, $b_i(\varepsilon)$ and $\nu_i(\varepsilon)$ such that

$$(8.10) \qquad \left| \sum_{i=1}^{n(\varepsilon)} b_i(\varepsilon) S\left(\frac{\nu_i(\varepsilon) L}{N(\varepsilon)}\right) \right| < \varepsilon \, , \quad \sum_{i=1}^{n(\varepsilon)} b_i(\varepsilon) = L \, .$$

From this sequence of values we select a subsequence such that the limit

$$\lim n(\varepsilon) = n$$

$$\lim \frac{\nu_i(\varepsilon) L}{N(\varepsilon)} = s_i \qquad , \quad i = 1,2,\dots,n$$

$$\lim b_i(\varepsilon) = b_i \qquad , \quad i = 1,2,\dots,n$$

shall exist as $\varepsilon \to 0$ through this particular subsequence. The existence of such a subsequence follows from the local compactness of finite dimensional spaces. Passing to the limit in (8.10) we obtain

$$(8.11) \qquad \sum_{i=1}^{n} b_i S(s_i) = 0 \, , \quad \sum_{i=1}^{n} b_i = L \, .$$

We introduce the abbreviation $S(s_i) = S_i$, $i = 1,2,\dots,n$.

From the minimal property of minor arcs of great circles it follows that

$$(8.12) \qquad \ell = \text{length of } S(s) \geq \sum_{i=1}^{n} |\widehat{S_i S_{i+1}}| \quad (S_{n+1} = S_1)$$

where $|\widehat{S_i S_j}|$ denotes the length of the minor arc $\widehat{S_i S_j}$ connecting S_i and S_j.

We shall show that there exist on the arcs $\widehat{S_i S_{i+1}}$, $i = 1,2,\dots,n$ two antipodic points (i.e. two points on the opposite ends of a diameter of the unit sphere).

The necessary and sufficient condition for two points S and S' to be antipodic is that a relation of the form

$$(8.13) \qquad bS + b'S' = 0 \, , \quad b > 0 \, , \quad b' > 0$$

be satisfied.

We distinguish 3 cases.

(a) $n = 2$.

In this case it follows from (8.11) and (8.13) that S_1 and S_2 are antipodic.

(b) $n = 3$.

In this case it follows from (8.11) that the points S_1, S_2, S_3 lie on a great circle made up of the arcs $\widehat{S_1 S_2}$, $\widehat{S_2 S_3}$, $\widehat{S_3 S_1}$; consequently the antipode of S_1 will lie on $\widehat{S_2S_3}$.

(c) $n = 4$.

In this case we define

$$b_{12} = |b_1S_1 + b_2S_2|\ ,\ b_{34} = |b_3S_3 + b_4S_4| \ .$$

($|X|$ stands for the distance of X from O , the center of the unit sphere).

If $b_{12} = 0$ (or $b_{34} = 0$), S_1 and S_2 (or S_3 and S_4) satisfy a relation of the form (8.13), therefore they are antipodic.

If $b_{12}, b_{34} > 0$, we define

$$S_{12} = \frac{1}{b_{12}}(b_1S_1 + b_2S_2)\ ,\ S_{34} = \frac{1}{b_{34}}(b_3S_3 + b_4S_4)\ ; \tag{8.14}$$

substituting this into (8.11) we obtain

$$b_{12}S_{12} + b_{34}S_{34} = 0$$

which is a relation of the form (8.13), hence S_{12} and S_{34} are antipodic. But it follows from (8.14) that S_{12} is on $\widehat{S_1S_2}$, S_{34} on $\widehat{S_3S_4}$.

This completes our demonstration that there always exist antipodic points S and S' on the arcs $\widehat{S_iS_{i+1}}$, $i = 1,2,\ldots,n$. Then from the minimal property of minor arcs of great circles it follows that

$$2\pi = |SS'| + |S'S| \leq \sum_{i=1}^{n} |\widehat{S_iS_{i+1}}|$$

which by (8.12) is $\leq \ell$.

Q.e.d.

It can be easily verified that the sign of equality holds for and only for plane convex curves.

CHAPTER III

The Isoperimetric Inequality and Related Inequalities

Section 1.

In this chapter the isoperimetric inequality and related inequalities will be discussed.

The isoperimetric inequality states that the area enclosed by a simple closed curve C is $\leq$ the area of the circle with the same circumference, the two areas being equal if and only if C is a circle.

There are numerous geometrical proofs of the isoperimetric inequality varying in elegance and simplicity. Of the analytical proofs, the first one was given by Hurwitz in 1901. We shall discuss it later. We shall first consider a strikingly simple demonstration due to E. Schmidt (1939).

Let C be a simple closed plane curve possessing a continuous tangent, which is cut by any straight line at most a finite number of times. Let A denote the area, L the total arc length of C.

We represent C parametrically by $\{x(t), y(t)\}$, $t_o \leq t \leq t_1$, $\{x(t_o), y(t_o)\} = \{x(t_1), y(t_1)\}$. The formulas

$$A = \int_{t_o}^{t_1} xy' dt \tag{1.1}$$

$$A = -\int_{t_o}^{t_1} yx' dt , \tag{1.2}$$

$$(y' = dy/dt, \ x' = dx/dt)$$

hold for all parametric representations where the variation of the arc length s, measured from $\{x(t_o), y(t_o)\}$, as t goes from t_o to t_1 is equal to L. This amounts to saying that $\{x(t), y(t)\}$ goes around C just once as t goes from t_o to t_1. It is important to note that it is not required that the representation $\{x(t), y(t)\}$ be one-to-one.

We enclose C between two vertical supporting lines touching the curve in P and Q respectively as shown on the following diagram. We draw a circle $\bar{C}$ having the same two vertical supporting lines. Let the radius of $\bar{C}$ be ϱ and the center O of $\bar{C}$ be the origin of our coordinate system. Let s be the arc length measured counterclockwise from P , $C = (x(s),y(s))$; then $Q = (x(s_o),y(s_o))$ for some s_o .

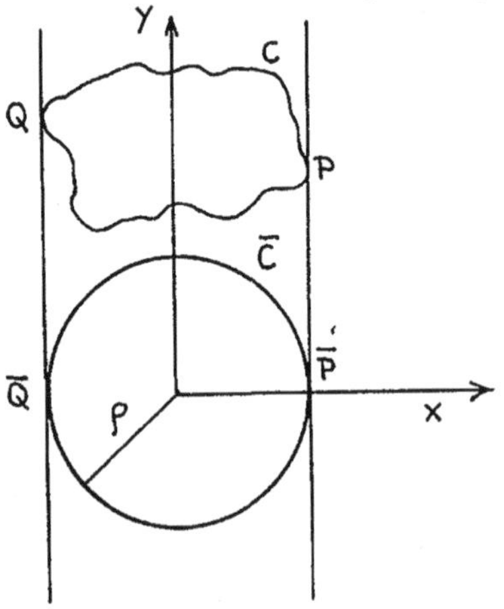

We introduce the following parametric representation for $\bar{C} = (\bar{x}(s),\bar{y}(s))$:

(1.3)
$$\begin{cases} \bar{x}(s) = x(s) \\ \bar{y}(s) = \sqrt{\varrho^2-x^2(s)} , & 0 \leq s \leq s_o \\ \quad\;\; = -\sqrt{\varrho^2-x^2(s)} , & s_o \leq s \leq L \end{cases}$$

(This parametric representation amounts to coordinating points of C to those of $\bar{C}$ by vertical projection of the arcs PQ, QP of C on the arcs $\overline{PQ}$, $\overline{QP}$ of $\bar{C}$ respectively; it is easy to see that as s varies from 0 to L , $(\bar{x}(s),\bar{y}(s))$ goes around $\bar{C}$ once. We apply (1.1) to C , (1.2) to $\bar{C}$:

$$A = \int_0^L xy'ds \; , \quad \bar{A} = \varrho^2\pi = -\int_0^L \bar{y}\bar{x}'ds = -\int_0^L \bar{y}x'ds \; .$$

Adding these two expressions we obtain

(1.4)
$$A + \varrho^2\pi = \int_0^L (xy'-\bar{y}x')ds \leq \int_0^L \sqrt{x^2+\bar{y}^2}\,\sqrt{x'^2+y'^2}ds$$

where the inequality on the right of (1.4) was obtained by applying Cauchy's inequality for the integrand. But $x'^2+y'^2 = 1$ since s is the arc length on C , and $x^2+\bar{y}^2 = \varrho^2$ by (1.3). Hence (1.4) gives

(1.5)
$$A + \varrho^2\pi \leq \int_0^L \varrho\,ds = L\varrho \; .$$

Applying the inequality between the arithmetic and the geometric mean of two numbers A and $\varrho^2\pi$ we obtain from (1.5)

$$2\sqrt{\varrho^2\pi A} = 2\varrho\sqrt{\pi A} \leq A + \varrho^2\pi \leq L\varrho \; ,$$

and after dividing by ϱ and squaring both sides we obtain

$$(1.6) \qquad 4\pi A \leq L^2$$

which is the celebrated isoperimetric inequality.

We shall show now that the sign of equality can hold if and only if C is a circle:

Assume that the sign of equality holds in (1.6); then the arithmetic and geometric means of A and $\varrho^2\pi$ must be equal, which is the case if and only if $A = \varrho^2\pi$; but since the choice of the y direction is arbitrary, this implies that the width (2ϱ) of C is a constant for all directions.

For the sign of equality to hold in (1.4) we have to have $(x,\bar{y})$ proportional to $(y'-x')$ with ϱ as the constant of proportionality, i.e. $x = \varrho y'$, $\bar{y} = -\varrho x'$; squaring the first of these equations we obtain $x^2 = \varrho^2 y'^2$, and by interchanging the x and y axes and using the fact that ϱ is independent of the direction of the coordinate axes we obtain $y^2 = \varrho^2 x'^2$. Adding these equations we obtain

$$x^2 + y^2 = \varrho^2(x'^2 + y'^2) = \varrho^2 ,$$

which means that C is situated on a circle of radius ϱ , hence it is identical with the circle.

Q.e.d.

Section 2. Generalization to n dimensions.

The generalization of the isoperimetric inequality to n dimension is an estimate of the volume A enclosed by a closed surface in terms of its surface area L . Since we expect the sign of equality to hold for and only for n dimensional spheres, and since for an n dimensional sphere of radius r , $A = C_n r^n$, $L = nC_n r^{n-1}$ (C_n is the volume of the n-dimensional sphere of unit radius), we conjecture the inequality

$$(2.1) \qquad C_n n^n A^{n-1} \leq L^n .$$

We shall prove (2.1) for the special class of bodies whose orthogonal projection on an $(n-1)$ dimensional plane is an $(n-1)$ dimensional sphere, by methods similar to those used in the previous paragraph.

Let S be the surface of an n-dimensional body with a continuously turning tangent plane whose projection on the plane formed by the $x_1, x_2, \ldots, x_{n-1}$ axes is an $n-1$ dimensional sphere; we assume furthermore that any straight line parallel to any of the axes cuts S a finite number of times. Denote the direction cosines of the normal at any point of S by $\cos\xi_\nu$, $\nu = 1,2,\dots,n$. Then, under the conditions imposed on S it is easy to show that

$$(2.2) \qquad A = \int_S x_\nu \cos\xi_\nu ds , \quad \nu = 1,2,\dots,n$$

where the surface integral is extended over the surface S.

We construct an n-dimensional sphere $\bar{S}$ whose projection on the hyper plane of the $x_1,x_2,\dots,x_{n-1}$ axes coincides with that of S. The center of $\bar{S}$ is chosen as origin of the coordinate system. Then

$$(2.3) \qquad C_n\varrho^n = \bar{A} = \int_S \bar{x}_n \cos\xi_n ds ,$$

where $\bar{A}$ denotes the volume of $\bar{S}$ and ϱ its radius.

We substitute $\nu = 1,2,\dots,n-1$ into (2.2) and add these equations to (2.3); we obtain:

$$(n-1)A + C_n\varrho^n = \int_S (x_1\cos\xi_1 + x_2\cos\xi_2 + \dots + x_{n-1}\cos\xi_{n-1} + \bar{x}_n\cos\xi_n) ds ;$$

estimating the integral on the right side by Schwarz's inequality we obtain

$$(2.4) \qquad (n-1)A + C_n\varrho^n \leq \int \sqrt{(x_1^2 + x_2^2 + \dots + x_{n-1}^2 + \bar{x}_n^2)} \cdot \sqrt{(\cos^2\xi_1 + \cos^2\xi_2 + \dots + \cos^2\xi_\nu)} ds = \int \sqrt{\varrho^2} ds = \varrho L .$$

Since the arithmetic mean of n numbers is $\geqslant$ their geometric mean, the left side of (2.4) is $\geqslant n\sqrt[n]{C_n A^{n-1}\rho^n}$, hence by (2.4)

$$n\sqrt[n]{C_n A^{n-1}\rho^n} \leq \rho L ;$$

raising both sides to the nth power we obtain (2.1). By a reasoning similar to the one used in the two-dimensional case we can show that the sign of equality holds if and only if S is the surface of an n-dimensional sphere.

Q.e.d.

Section 3.

We shall now consider the older proof of Hurwitz for the isoperimetric inequality in two dimensions. Our presentation is the one given in Hardy-Littlewood-Pólya's "Inequalities".

Lemma I. (Wirtinger's inequality): If $f(t)$ is a continuous function of period 2π, possessing a continuous derivative $f'(t)$ (*), and $\int_0^{2\pi} f(t)\,dt = 0$; then

$$\int_0^{2\pi} f'^2 dt \geq \int_0^{2\pi} f^2 dt \,, \tag{3.1}$$

the sign of equality holding if and only if $f(t) = a \cos t + b \sin t$. The condition $\int f(t)\,dt = 0$ is not superfluous since otherwise we could make the right side of (3.1) arbitrarily large without altering the left side, by adding any constant to $f(t)$.

Hurwitz in his original proof resorts to the theory of Fourier series; namely $f(t)$ and $f'(t)$ both being continuous the Fourier series of the latter is the term-by-term derivative of that of $f(t)$:

$$\begin{aligned} \pi f(t) &\sim a_0/2 + \sum_{n=1}^{\infty} (a_n \cos nt + b_n \sin nt) \\ \pi f'(t) &\sim \Sigma (nb_n \cos nt - na_n \sin nt) \end{aligned} \tag{3.2}$$

hold. Since $a_0 = \int_0^{2\pi} f(t)\,dt$, it follows from the hypothesis that $a_0 = 0$.

Applying Parseval's formula to the Fourier expansions (3.2) we obtain

$$\int f^2 dt = \sum_{n=1}^{\infty} (a_n^2 + b_n^2)$$

$$\int f'^2 dt = \sum_{n=1}^{\infty} n^2 (a_n^2 + b_n^2) \,.$$

Then

$$\int f'^2 dt - \int f^2 dt = \sum_{n=1}^{\infty} (n^2-1)(a_n^2 + b_n^2) \,,$$

and this expression is always ≥ 0. Its value is zero if and only if $a_n = b_n = 0$ for all $n > 1$, i.e. $f = a_1 \cos t + b_1 \sin t$. This proves Lemma 1.

Q.e.d.

Proof of the isoperimetric inequality: Let C be a simple closed curve with piecewise continuous tangent; we denote its area by A and its total arc length by L. Without loss of generality we can take $L = 2\pi$.

We choose our rectangular coordinate system so that the center of gravity of the circumference falls on the y axis, i.e.

$$\int_0^{2\pi} x \, ds = 0$$

(*) It is enough to assume that $f'(t)$ is square integrable and $f(t)$ can be represented as the integral of its derivative.

where the parameter s is the arc length. By (1.1)

(1.1) $$A = \int xy' ds \ .$$

Also, since $x'^2 + y'^2 = 1$,

(3.3) $$2\pi = \int_0^{2\pi} (x'^2 + y'^2) ds \ .$$

We multiply (1.1) by 2 and subtract it from (3.3):

(3.4) $$\begin{aligned} 2(\pi - A) &= \int_0^{2\pi} (x'^2 + y'^2 - 2xy') \, ds \\ &= \int_0^{2\pi} (x'^2 - x^2) ds + \int_0^{2\pi} (x - y')^2 ds \ . \end{aligned}$$

The first term on the right side is $\geqslant 0$ by Wirtinger's inequality; the second term is $\geqslant 0$ because it is the integral of a non-negative quantity; hence

(3.5) $$2(\pi - A) \geqslant 0 \quad \text{or} \quad A \leqslant \pi$$

i.e. the area enclosed by the curve C having the circumference 2π is $\leqslant \pi$ = the area of the circle having the circumference 2π. If the sign of equality holds in (3.5), it must hold in Lemma I as applied to $f = x(s)$, which is the case if and only if $x(s) = a \cos s + b \sin s$; in addition $x - y' \equiv 0$ must hold, i.e. $y = a \sin s - b \cos s + c$; it is easy to see that this is a parametric representation of a circle.

This proof of the isoperimetric inequality is not as elementary as the one given in section 1 since in the proof of Lemma I it makes use of the more sophisticated theory of Fourier series. It is therefore desirable to find an elementary proof of Wirtinger's inequality; such a proof is suggested by the procedure of section 1. (*)

If $f(t)$ satisfies the condition of $\int_0^{2\pi} f(t) dt = 0$, it follows for $\bar{f}(t) = f(t) + C$

(3.6) $$\int_0^{2\pi} \bar{f}^2(t) dt = \int_0^{2\pi} f^2 dt + 2\pi C^2 \ .$$

If $M \geqslant 0$ denotes $\max\limits_{0 \leqslant t \leqslant 2\pi} f(t)$, $m \leqslant 0$ denotes $\min\limits_{0 \leqslant t \leqslant 2\pi} f(t)$,

---

(*) For yet another proof see Hardy-Littlewood-Pòlya "Inequalities".

the function

$$x(t) = f(t) - \frac{M+m}{2}$$

will have the property $\max x(t) = -\min x(t) = (\frac{M-m}{2}) = \varrho$. Assume that a maximum occurs at $t = 0$, a minimum at $t = t_o$, $0 < t_o < 2\pi$.

Construct in the x,y plane a circle of radius ϱ . We introduce t as a parameter for the circle by the equation

$$\bar{x}(t) = x(t)$$
$$\bar{y}(t) = \sqrt{\varrho^2 - x^2(t)}\ , \quad 0 \leq t \leq t_o$$
$$\bar{y}(t) = -\sqrt{\varrho^2 - x^2(t)}\ , \quad t_o \leq t \leq 2\pi\ .$$

It can be easily verified that as t goes from 0 to 2π , the point $(\bar{x}(t), \bar{y}(t))$ goes around the circle once.

If we assume for sake of simplicity that $x(t)$ takes up the same value in its range for a finite number of different values of t , then it is easy to show that the area of the circle is given by the well known formula

$$\varrho^2\pi = -\int_0^{2\pi} \bar{y}\bar{x}'dt\ . \tag{3.7}$$

We estimate the right side of (3.7) by Schwarz's inequality: we introduce the abbreviation $\int_0^{2\pi} x^2dt = A$ $\int_0^{2\pi} x'^2dt = B$:

$$\varrho^2\pi \leq \sqrt{\int \bar{y}^2dt \int \bar{x}'^2dt} = \sqrt{\int (\varrho^2 - x^2)dt \int x'^2dt} = \sqrt{(2\pi\varrho^2 - A)B}\ .$$

Multiplying both sides by $\sqrt{A}$ we obtain

$$\varrho^2\pi\sqrt{A} \leq \sqrt{A(2\pi\varrho^2 - A)B} \tag{3.8}$$

and since the product $A(2\pi\varrho^2 - A)$, being maximum for $A = \pi\varrho^2$, is always $\leq \pi^2\varrho^4$, (3.8) yields after division by $\varrho^2\pi$

$$\sqrt{A} \leq \sqrt{B}\ ,\ \text{i.e.}\ \ A \leq B \ \ \text{or}\ \ \int_0^{2\pi} x^2dt \leq \int_0^{2\pi} (x')^2dt \tag{3.9}$$

Q.e.d.

Discussion of the sign of equality: If the sign of equality holds in (3.9), then it must hold in Schwartz's inequality as applied to (3.7), i.e.

$$\bar{y} = k\bar{x}' = kx' \tag{3.10}$$

Also

$$\int_0^{2\pi} \bar{y}^2dt = \int_0^{2\pi} (\varrho^2 - x^2)dt = 2\pi\varrho^2 - A = \pi\varrho^2\ ,$$

since $A(2\pi\varrho^2 - A) = \pi^2\varrho^4$ must hold if in (3.9) the sign of equality holds. By (3.10)

$$\int \bar{y}^2 dt = \pi\varrho^2 = k^2\int x'^2 dt = k^2 B = k^2\pi\varrho^2 ,$$

hence $k^2 = 1$.

Squaring (3.10) and substituting $\bar{y}^2 = \varrho^2 - x^2$ we obtain for x the differential equation

$$x'^2 = \varrho^2 - x^2 , \quad \text{i.e.} \quad x = a \cos t + b \sin t$$

as a necessary (and sufficient) condition for the sign of equality to hold.

Substituting (3.9) into (3.6) we obtain

$$\int_0^{2\pi} f^2 dt \leq \int_0^{2\pi} f'^2 dt - \left(\frac{M+m}{2}\right)^2 2\pi$$

which is a slightly sharper inequality than (3.1).

Section 4.

This method of proof of Wirtinger's inequality can be applied to the following more general type of inequality: Let $f(t)$ be a function defined for $0 \leq t \leq 1$ having a continuous derivative in this interval which satisfies the condition

$$\max f(t) = f(0) = -\min f(t) = -f(t_0) = \varrho .$$

Let a, b be two numbers satisfying the conditions $a > 0$, $b > 1$; define β by the equation $1/b + 1/\beta = 1$. Then

$$\left(\int_0^1 |f|^a dt\right)^{1/a} \leq c_{ab}\left(\int_0^1 |f'|^b dt\right)^{1/b} \tag{4.1}$$

where

$$c_{ab} = \frac{a}{4B(1/\beta+1, 1/a)} \left(\frac{\beta}{\beta+a}\right)^{1/a} \left(\frac{a}{\beta+a}\right)^{1/\beta}$$

($B(p,g)$ is the Beta function.) The sign of equality holds for the function $f(t)$ which satisfies the differential equation

$$\frac{\rho^{a-b} a^b}{(4B(1/\beta+1, 1/a))^b} \left(\frac{a}{a+\beta}\right)^b |f'(t)|^b = \varrho^a - |f(t)|^a . \tag{4.2}$$

Proof: For the curve defined by

$$|x|^a + |y|^\beta = \varrho^a \tag{4.3}$$

we introduce the parametric representation

$$x = f(t) \ ,$$

$$y = \sqrt[\beta]{\varrho^a - |f(t)|^a} \qquad \text{for} \quad 0 \leq t \leq t_o$$

$$-\sqrt[\beta]{\varrho^a - |f(t)|^a} \qquad \text{for} \quad t_o \leq t \leq 1 \ .$$

The curve (4.3) encloses an area $= u\varrho^{1+a/\beta}$ where $u = 4/aB(1/\beta+1, 1/a)$. By formula (1.2) for the area enclosed by a curve:

$$u\varrho^{1+a/\beta} = -\int_0^1 yx'dt \tag{4.4}$$

we estimate the right hand side of (4.4) by Hölder's inequality; introducing the abbreviation,

$$\left(\int_0^1 |f(t)|^a dt\right)^{1/a} = A \ , \ \left(\int_0^1 |f'(t)|^b\right)^{1/b} = B \ :$$

$$u\varrho^{1+a/\beta} \leq \left(\int_0^1 y^\beta dt\right)^{1/\beta} \left(\int_0^1 x'^b dt\right)^{1/b} = (\varrho^a - A^a)^{1/\beta} \cdot B \ . \tag{4.5}$$

We multiply both sides of (4.5) by A and replace the factor $(A^\beta(\varrho^a - A^a))^{1/\beta}$ by its maximum value $= \varrho^{1+a/\beta}(\beta/\beta+a)^{1/a}(a/\beta+a)^{1/\beta}$, assumed for $A = \varrho(\beta/\beta+a)^{1/a}$:

$$Au\varrho^{1+a/\beta} \leq \varrho^{1+a/\beta}(\beta/\beta+a)^{1/a}(a/a+\beta)^{1/\beta}B \ .$$

Dividing by $u\varrho^{1+a/\beta}$ we obtain (4.1). The sign of equality holds if and only if:

(a) In Hölder's inequality as applied to (4.4) the sign of equality holds;

(b) $A = \varrho(\beta/\beta+a)^{1/a}$.

From (a) it follows that

$$|y|^\beta = k|x'|^b \tag{4.6}$$

for some constant k ; by integrating (4.6) and using condition (b) we find that k has the value $\varrho^{a-b}(a/a+\beta)^b/u^b$; substituting this value of k into (4.6) we obtain the differential equation (4.2). The demonstration that (4.2) has a continuous solution for $0 \leq t \leq 1$ whose maximum and minimum are respectively $+\varrho$ and $-\varrho$ will be omitted. (*)

(*) Compare E. Schmidt "Ueber die Ungleichung welche die Integrale über eine Potenz einer Funktion und über eine andere Potenz ihrer Ableitung verbindet", Math. Ann. 117, 301-326, (1940).

Section 5.

In this paragraph we shall complete the proof of the three-dimensional isoperimetric inequality by constructing to an arbitrary body of surface area L and volume A a rotational body of the same surface area and greater volume, or, what is equivalent, a rotational body of the same volume and lesser surface area. (*)

We assume that all surfaces considered in this paragraph have continuous derivatives.

Lemma II:

$$\int_0^T \sqrt{1+D^2(t)}\,dt \geq \sqrt{T^2+\left(\int_0^T D\,dt\right)^2} \tag{5.1}$$

holds for all piecewise continuous functions $D(t)$, the sign of equality holding if and only if $D(t)$ is a constant.

Proof: If $x(t)$, $y(t)$, $0 \leq t \leq T$ are two arbitrary functions with piecewise continuous derivatives $x'(t) = f(t)$, $y'(t) = g(t)$, the arc $[(x(t),y(t)]$ connecting the points $[x(0),y(0)]$ and $(x(T),y(T))$ is rectifiable and its length ℓ =

$$\int_0^T \sqrt{x'^2+y'^2}\,dt = \int_0^T \sqrt{f^2+g^2}\,dt .$$

The length of the chord connecting the same two points

$$= \sqrt{(x(T)-x(0))^2+(y(T)-y(0))^2} = \sqrt{\left(\int_0^T f(t)\,dt\right)^2+\left(\int_0^T g(t)\,dt\right)^2} ,$$

hence the inequality

$$\int_0^T \sqrt{f^2+g^2}\,dt \geq \sqrt{\left(\int_0^T f\,dt\right)^2+\left(\int_0^T g\,dt\right)^2} \tag{5.2}$$

holds for all piecewise continuous functions f and g; by putting $f(t) = 1$, $g(t) = D(t)$ we obtain (5.1) as a special case of (5.2). If we are given any three-dimensional body we consider its intersection with the planes z = const. The arc length σ of the boundary of this intersection and z are now taken as parameters for the surface of the body.(**) If $D(z,\sigma)$ denotes the Jacobian $\frac{\partial(x,y)}{\partial(z,\sigma)}$, the surface element is

$$dL = \sqrt{1+D^2}\,dz\,d\sigma .$$

(*) The following proof is due to H.A. Schwarz (1884).

(**) We assume that the only horizontal tangent planes are the two supporting planes; otherwise difficulties arise as z and σ are introduced as parameters.

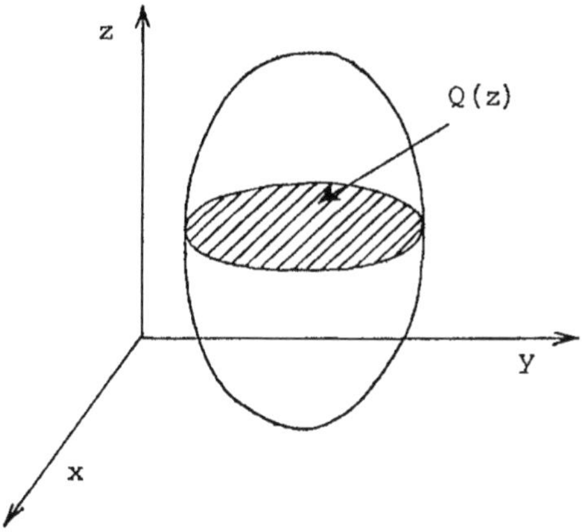

Denote the area of the intersection of the plane parallel to the (x,y) plane at height z with the body by $Q(z)$; since $\frac{\partial(x,y)}{\partial(z,\sigma)}\,dz d\sigma$ is the infinitesimal area of the projection of the surface element dL on the x,y plane, $\frac{dQ}{dz} = \int_0^{\ell(z)} D d\sigma$, where $\ell(z)$ is the circumference of $Q(z)$.

In the formula
$$L = \int dz \int_0^{\ell(z)} \sqrt{1+D^2}\, d\sigma$$
we estimate the integral with respect to $d\sigma$ by Lemma II:

$$(5.3) \qquad L \geq \int \sqrt{\ell^2(z) + \left(\int_0^{\ell(z)} D d\sigma\right)^2}\, dz = \int \sqrt{\ell^2(z) + \left(\frac{dQ}{dz}\right)^2}\, dz$$

In (5.3) the sign of equality holds if D does not depend upon σ ; this will certainly be the case for rotational surfaces.

Given an arbitrary body F of surface area L and volume A and cross sectional area $Q(z)$ for any z , we construct another body F' whose horizontal cross section at height z is a circle of area $Q(z)$ having its center on the z-axis. Let A' and L' denote the volume and surface area of F' . By cavalieri's principle $A' = A$. Since F' is a rotational body, in (5.3) as applied to F' the sign of equality holds:

$$(5.3') \qquad L = \int \sqrt{(\ell'(z))^2 + \left(\frac{dQ}{dz}\right)^2}\, dz$$

where $\ell'(z)$ denotes the circumference of the circle of area $Q(z)$ (i.e. $\ell'(z) = \sqrt{4\pi Q(z)}$) . The isoperimetric inequality for <u>two</u> dimensions asserts that

$$(5.4) \qquad \ell'(z) \leq \ell(z) \ .$$

Combining (5.3), (5.3') and (5.4) we obtain $L \geq L'$, with the sign of equality holding if and only if F has rotational symmetry about an axis parallel to the z axis. Q.e.d.

CHAPTER IV

The Elementary Concept of Area and Volume

Section 1.

In elementary geometry area (volume) is defined in the following manner:

(i) A rectangle (right rectangular prism) with sidelengths $1, \ell$, $(1,1,\ell)$ units is called a normal rectangle with an area (volume) of ℓ square (cubic) units.

(ii) The area (volume) of any polygon (polyhedra) P which can be decomposed into a finite number of parts so that these parts can be rearranged to form a normal rectangle (rectangular prism) R is equal to the area (volume) of R .

This definition, in order that it be useful, must be:

(a) consistent

(b) applicable to a sufficiently wide class of (preferably all) polygons (polyhedra) P .

The consistency of (i) and (ii) means that if a polygon can be decomposed and rearranged into two normal rectangles R and R' then $R \cong R'$. This consistency could be demonstrated,(*) for example, by means of Jordan measure theory.

In this chapter we shall investigate point sets to which this definition is applicable. We shall start our investigations by scrutinizing the derivations of the formulas for the area of a triangle and the volume of a tetrahedron.

Given a triangle (abc) , denote by h the length of the altitude to side ab , by ℓ the length of side ab , and by A the area of the triangle. Then

(1.1) $$A = \frac{1}{2} h\ell$$

Proof: Decompose and rearrange the triangle as shown on the accompanying diagram, to form a rectangle (a b b'a') of sidelengths ℓ and h/2 .

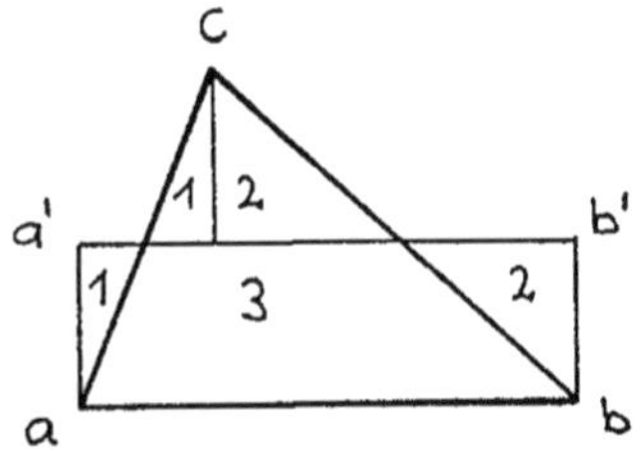

(*) For a systematic treatment see: D. Hilbert, "Grundlagen der Geometrie", Chapter IV.

If we assume - and this assumption will be proved in section 2, Lemma II - that the area of <u>any</u> rectangle is the product of the lengths of its sides, then this rearrangement proves (1.1).

Given a tetrahedron (1 2 3 1') , denote by h the length of the altitude to the face (1 2 3) whose area we denote by A ; V denotes the volume of (1 2 3 1') .

(1.2) $V = 1/3\ hA$.

<u>Proof</u>: Construct the triangular prism (1 2 3 1'2'3') ; assume - and this assumption will be proved in section 6, Theorem V - that the volume of the prism is = base area times altitude = hA .

The prism is the sum of three tetrahedra: (1 2 3 1') , (1'2 3 2') and (1'2'3'3) ; any two of these tetrahedra have congruent faces

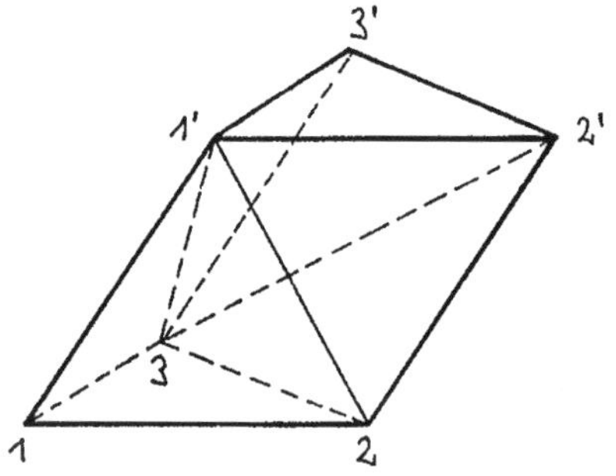

with equal altitudes to these faces. (E.g. (1 2 3 1') and (1'2 3 2') have the congruent faces (1 1'2) and (1'2'2) with the opposite vertex 3 in common.)

If we assume that

(iii) Two tetrahedra with the same base and equal altitudes have the same volume.

Then it follows that the three tetrahedra (1 2 3 1') , (1'2 3 2') and (1'2'3'3) have the same volume = one third of the volume of the prism = 1/3 h A . Q.e.d.

In this proof in addition to (i) and (ii) we had to use postulate (iii); therefore in view of our present definition of volume this proof is invalid unless (iii) can be derived from (i) and (ii). Whether this is possible or not was for a long time one of the outstanding unsolved problems of geometry, proposed by Gauss and solved for the first time by M. Dehn (*) in the negative.

(*) M. Dehn, Math. Ann., 55, 465-478 (1902).

Section 2. Definition:

Two polygons (polyhedra), A and B, are equivalent if there exist polygons (polyhedra) A_i, B_i, $i = 1,2,\dots,n$ such that

$$(2.1)\qquad A = \sum_{i=1}^{n} A_i \qquad B = \sum_{i=1}^{n} B_i$$

$$(2.2)\qquad A_i \cong B_i \ , \quad i = 1,2,\dots,n$$

and A_i, A_j, and B_i, B_j, have no interior point in common for $i,j = 1,2,\dots,n$ $i \neq j$. Equivalence will be denoted by $A \sim B$.

Remark: The elementary definition of area (volume) is applicable to a point set P if and only if $P \sim R$, where R is a normal rectangle (rectangular prism). In all subsequent paragraphs, unless otherwise specified, the words area and volume will be used in the Jordan sense and the elementary notion of area and volume will be replaced by the idea of equivalence.

Lemma I: Equivalence is transitive, i.e. $A \sim B$ and $B \sim C$ imply $A \sim C$.

Proof: By hypothesis $A \sim B$ and $B \sim C$. Therefore there exist A_i, B_i, $i = 1,2,\dots,m$ and B'_j, C_j, $j = 1,2,\dots,n$ such that no two A_i, B_i or B'_j, C_j have an interior point in common and

$$A = \sum_{i=1}^{m} A_i \ , \qquad B = \sum_{i=1}^{m} B_i$$

$$B = \sum_{j=1}^{n} B'_j \ , \qquad C = \sum_{j=1}^{n} C_j$$

$$A_i \cong B_i \ , \quad i = 1,2,\dots,m$$

$$B'_j \cong C_j \ , \quad j = 1,2,\dots,n \ .$$

We define S_{ij} as the common part of B_i and B'_j, $i = 1,2,\dots,m$, $j = 1,2,\dots,n$. Since B_i and B'_j are themselves polygons (polyhedra), so are the S_{ij}. All common points of S_{ij} and $S_{k\ell}$ are common points of B_i, B_k, B'_j, B'_ℓ; since the first two and the last two polygons (polyhedra) have inner points in common only if $i = k$, $j = \ell$, it follows that S_{ij} and $S_{k\ell}$ have no interior points in common unless $i = k$, $j = \ell$.

From the definition of S_{ij} it follows that

$$(2.3)\qquad \begin{aligned} B_i &= \sum_{j=1}^{n} S_{ij} , \quad i = 1,2,\ldots,m \\ B'_j &= \sum_{i=1}^{m} S_{ij} , \quad j = 1,2,\ldots,n . \end{aligned}$$

Then, since $A_i \cong B_i$ and $C_j \cong B'_j$, we can carry the subdivision (2.3) of B_i over to A_i and similarly that of B'_j over to C_j , i.e. there exist R_{ij} and T_{ij} , $i = 1,2,\ldots,m$, $j = 1,2,\ldots,n$ such that

$$(2.4)\qquad R_{ij} \cong T_{ij} \cong S_{ij} ,$$

$$(2.5)\qquad A_i = \sum_{j=1}^{n} R_{ij} , \quad i = 1,2,\ldots,m$$

$$(2.6)\qquad C_j = \sum_{i=1}^{m} T_{ij} , \quad j = 1,2,\ldots,n .$$

Summing (2.5) and (2.6) over all i and j respectively we find

$$(2.7)\qquad \begin{aligned} A &= \sum_{i=1}^{m} A_i = \sum_{i=1}^{m} \sum_{j=1}^{n} R_{ij} \\ C &= \sum_{j=1}^{n} C_j = \sum_{j=1}^{n} \sum_{i=1}^{m} T_{ij} \end{aligned}$$

(2.7) provides a subdivision of A and C into a finite number of polygons (polyhedra) which are pairwise congruent and such that no two of the R_{ij} or T_{ij} have an interior point in common (this follows from the similar property of the S_{ij}). Hence $A \sim C$. Q.e.d.

It follows from the transitivity of equivalence that we can put all equivalent polygons (polyhedra) into one class and thus divide all polygons (polyhedra) into equivalence classes.

<u>Theorem I</u>: All polygons with the same area belong to the same equivalence class.

<u>Corollary</u>: (i) and (ii) define an area in the elementary sense for every polygon.

<u>Lemma II</u>: Every polygon is equivalent to a normal rectangle.

<u>Proof of Lemma II</u>: Every polygon is the sum of a finite number of triangles. Since normal rectangles can be joined to form one normal rectangle, it is sufficient to prove Lemma II for triangles. This will be done in four steps:

a) Every triangle of altitude a and base b is equivalent to a parallelogram with altitude $a/2$ and base b .

This has already been demonstrated by the construction given in the proof for formula (1.1).

b) Two parallelograms which have one side and the altitude on this side in common are equivalent.

Let $(1\ 2\ 3\ 4)$ and $(1\ 2\ 3'4')$ be the two parallelograms. Since the altitudes on $(1\ 2)$ are the same in both parallelograms, the points $3, 4, 3'$ and $4'$ lie on a straight line. We distinguish two cases:

(A) $3'$ lies in the closed interval $(3,4)$. Then
$(1\ 2\ 3\ 4) = (1\ 3\ 3') + (1\ 3'4\ 2)$, $(1\ 2\ 3'4') = (2\ 4\ 4') + (1\ 3'4\ 2)$
and since $(1\ 3\ 3') \cong (2\ 4\ 4')$, we have $(1\ 2\ 3\ 4) \sim (1\ 2\ 3'4')$.

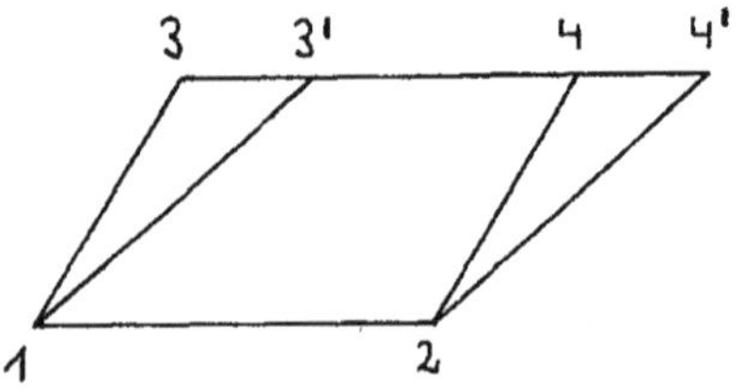

(B) $3'$ lies outside of $(3,4)$. We construct a finite sequence of points $3^{(1)}, 3^{(2)}, \ldots, 3^{(n)}$ on the line through 3 and $4'$ such that $3^{(i+1)}3^{(i)} = 3'4'$, $3^{(1)}$ coincides with $3'$, and $3^{(n)}$ lies in the closed interval $(3,4)$. The existence of such a sequence is guaranteed by the axiom of Archimedes.

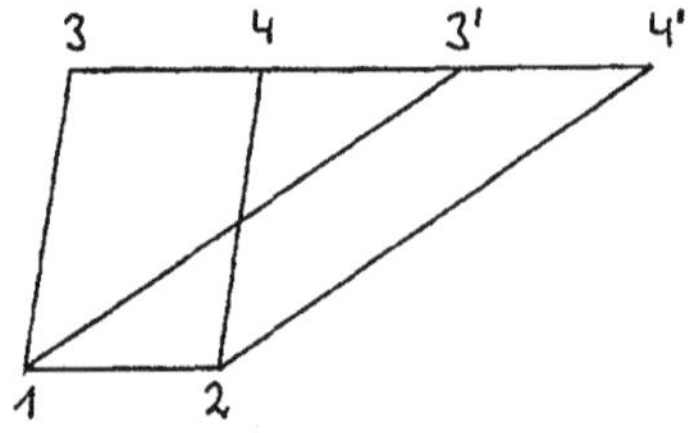

By (A) $(1\ 2\ 3^{(i)}3^{(i+1)}) \sim (1\ 2\ 3^{(i-1)}3^{(i)})$ for $i = 2,3,\ldots,n-1$ and also $(1\ 2\ 3^{(n-1)}3^{(n)}) \sim (1\ 2\ 3'4')$, and $(1\ 2\ 3^{(1)}3^{(2)}) \sim (1\ 2\ 3\ 4)$. Since equivalence is transitive, we have

$$(1\ 2\ 3\ 4) \sim (1\ 2\ 3'4') .$$

c) Two triangles with the same base and altitude are by a) both equivalent to two parallelograms with one side and an altitude in common. These parallelograms are, by b) equivalent, hence the transitivity of equivalence shows that the original triangles were also equivalent.

d) Given a <u>right</u> triangle (1 2 3) we select on the halfray {1,3} the point 3' whose distance from 1 is two units. We suppose {1,3} ⩾ {1,3'} although the case {1,3'} ⩾ {1,3} can be demonstrated in exactly the same manner. Then we select the point 2' so that the line through 3 and 2' is parallel to the line through 3' and 2 .

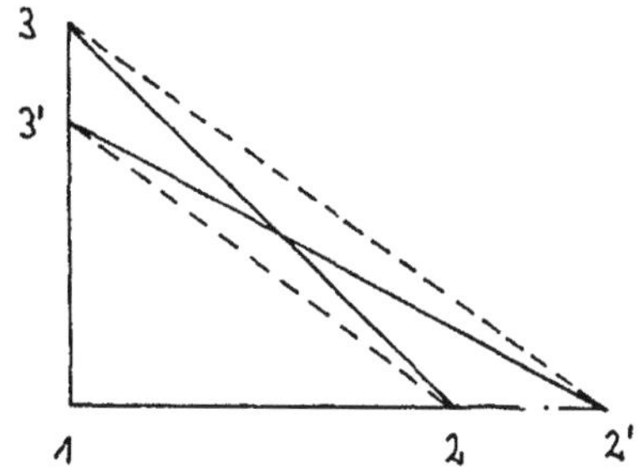

(2.8) (1 2'3') = (1 2 3') + (2 2'3')

and

(2.9) (1 2 3) = (1 2 3') + (2 3 3') ;

(2 2' 3') and (2 3 3') are triangles with the same base (2 3') and equal altitudes on this base; hence by c) they are equivalent and by (2.8) and (2.9)

(2.10) (1 2 3) ~ (1 2'3') .

But to any given triangle we can construct a right triangle (1 2 3) with the same base and altitude which, by c) will be equivalent to the original triangle. This, together with (2.10), shows that every triangle is equivalent to a right triangle one of whose legs is two units long. Combining this statement with a) we have completed our proof of Lemma II for triangles, and consequently for all polygons.

If P and P' are two polygons, then by Lemma II there exist two normal rectangles R and R' so that P ~ R , P' ~ R' . If P and P' have equal areas, so do R and R' , consequently R ≅ R' . Transitivity of equivalence shows then that P ~ P' . This completes the proof of Theorem I.

<u>Section 3</u>.

In the previous paragraph we have shown that if two polygons have the same area then both of them can be built up from the same finite collection of polygons. It is of some interest to find out (at least in some special cases) what is the least number of polygons that are needed.

Example: Pythagorean Theorem: If a, b, and c are the two legs and the hypotenuse of a right triangle respectively, then the area of the square with side-length c = the sum of the areas of the squares with sidelength a and b. The accompanying diagram shows how $a^2 + b^2$ and c^2 can be built up from the same five polygons. It can be shown that a subdivision into at least five parts is necessary for a demonstration of the Pythagorean theorem.

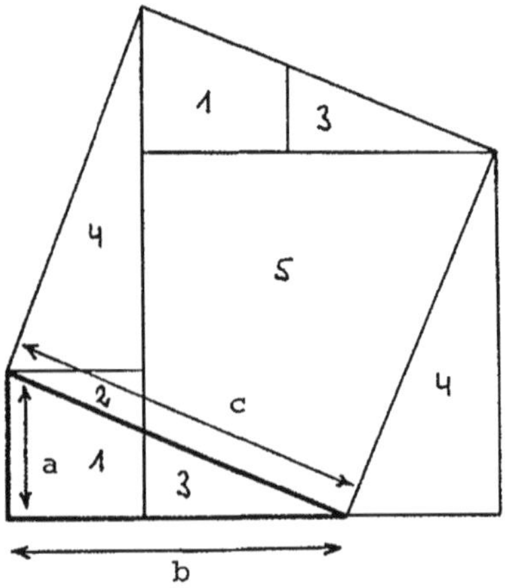

Section 4.

Before discussing the problem of equivalence of polyhedra we shall discuss the much simpler problem of equivalence with respect to regular subdivision.

Two polyhedra A and B are equivalent with respect to regular subdivision (denoted by $A \approx B$) if

$$A = \sum_{i=1}^{n} A_i \, , \; B = \sum_{i=1}^{n} B_i \, , \; A_i \cong B_i \, , \quad i = 1,2,\ldots,n \, ,$$ where A_i and B_i are polyhedra such that any two A_i or B_i either have no point in common or just one vertex, edge or face in common. (Compare this definition with the notion of subdivision in Chapter I.)

We have shown in Chapter I that two regular polyhedral subdivisions of a polyhedron always have a common regular polyhedral subdivision. From this it follows that $A \approx B$ and $B \approx C$ imply $A \approx C$, consequently we can define equivalence classes with respect to regular subdivision.

Lemma II. A necessary condition for $A \approx B$. Let A and B be two polyhedra and let α_i, $i = 1,2,\ldots,r$ and β_j, $j = 1,2,\ldots,s$, denote their dihedral angles. If $A \approx B$, then there exist positive integers m_i, n_j, $i = 1,2,\ldots,r$, $j = 1,2,\ldots,s$, and an integer k such that

$$\sum_{i=1}^{r} m_i\alpha_i = \sum_{j=1}^{s} n_j\beta_j + k\pi \quad . \tag{4.1}$$

Proof: Assume that $A \approx B$. Then there exist A_i, B_i, $A_i \cong B_i$, such that $A = \sum_{i=1}^{M} A_i$, $B = \sum_{i=1}^{M} B_i$ is a regular subdivision of A and B.

Denote the dihedral angles of the A_i, $i = 1,2,\ldots,M$, by φ_k, $k = 1,2,\ldots,N$.

We shall evaluate the sum $\sum_{i=1}^{N} \varphi_k$ by grouping together all dihedral angels φ_k that lie around a single edge of the sum ΣA_i. The sum of those φ_k that lie around an interior edge of the sum ΣA_i is 2π, while the sum of those φ_k which lie about a boundary edge is equal to the dihedral angle of that edge of A on which this boundary edge lies. Thus

$$\sum_{k=1}^{N} \varphi_k = \sum_{i=1}^{r} m_i\alpha_i + 2k_1\pi \ , \tag{4.2}$$

where m_i is the number of intervals into which the edge with the dihedral angle α_i is divided by the subdivision $A = \sum_{i=1}^{M} A_i$ and k_1 the number of interior edges in this subdivision.

Since $A_i \cong B_i$, $i = 1,2,\ldots,M$, the φ_k can be regarded as dihedral angles of the polyhedra B_i. Then by a reasoning identical to the one by which (4.2) was derived we derive

$$\sum_{k=1}^{N} \varphi_k = \sum_{i=1}^{s} n_j\beta_j + 2k_2\pi \ . \tag{4.3}$$

Therefore equating (4.2) and (4.3), and setting $k = 2k_2 - 2k_1$,

$$\sum_{i=1}^{r} m_i\alpha_i = \sum_{j=1}^{s} n_j\beta_j + k\pi \quad .$$

Q.e.d.

Theorem II: (*) There exist polyhedra with the same volume which are not equivalent with respect to regular subdivision.

(*) This theorem and proof are due to Bricard.

Proof: Let X denote the regular tetrahedron with unit volume and Y the unit cube. We shall prove Theorem II by showing that $X \not\approx Y$.

Let α_i , $i = 1,2,\ldots,6$, β_j , $j = 1,2,\ldots,12$, denote the dihedral angles of X and Y respectively. By an elementary calculation we have

$$\alpha_i = \gamma = \cos^{-1}1/3 \ , \qquad i = 1,2,\ldots,6 \ ,$$

$$\beta_j = \pi/2 \ , \qquad j = 1,2,\ldots,12 \ .$$

If $X \approx Y$ were true then by Lemma II

$$\sum_{i=1}^{6} m_i\alpha_i = \gamma \sum_{1}^{6} m_i = \sum n_j\beta_j + k\pi = \pi/2 \sum_{j=1}^{12} n_j + k\pi \tag{4.4}$$

would hold for some positive integers m_i, n_j and some integer k . We shall show that γ and π are incommensurable, hence a relation of the form (4.4) cannot hold, consequently $X \approx Y$ cannot hold either.

Define

$$\zeta = e^{i\gamma} = \cos\gamma + i\sin\gamma = 1/3 + \frac{i}{3}\sqrt{8} \ ;$$

ζ is the root of the following quadratic equation

$$3\zeta^2 - 2\zeta + 3 = 0 \ . \tag{4.5}$$

We shall first show that for all positive integer exponents m

$$3^{m-1}\zeta^m = a\zeta + b \ , \tag{4.6}$$

where a,b are integers depending on m satisfying the following condition:

$$a \not\equiv 0 \pmod 3 \ . \tag{4.7}$$

We prove this by induction.

(4.6) holds for $m = 2$ by virtue of (4.5). If we assume that (4.6) holds for m , then multiplying it by 3ζ and substituting for $3\zeta^2 = 2\zeta - 3$ we obtain

$$3^m\zeta^{m+1} = 3a\zeta^2 + 3b\zeta = a(2\zeta-3) + 3b\zeta = (2a+3b)\zeta - 3a \ .$$

Since a satisfies (4.7), the new coefficient $2a + 3b$ will evidently also satisfy (4.7) since $2a + 3b \not\equiv 0 \pmod 3$.

Suppose now to the contrary that γ is a rational multiple of π ; then ζ would be a root of unity, i.e. for some positive integer N

$$\zeta^N = 1$$

(4.6) yields for $m = N$

$$3^{N-1}\zeta^N = a\zeta + b = 3^{N-1} .$$

Equating the coefficients of the imaginary parts on both sides $a = 0$ follows, contrary to (4.7). Hence the assumption that γ is a rational multiple of π leads to a contradiction. This shows that the unit cube and unit regular tetrahedron are not equivalent with respect to regular subdivision.

Q.e.d.

Section 5.

In this paragraph we shall prove

Theorem III: There exist two polyhedra with the same volume which are not equivalent.

It follows from Theorem III - and this is the main problem to be discussed in this chapter - that it is not possible to develop an elementary theory of volume in three dimensions.

We shall prove Theorem III by a lemma analogous to Lemma II:

Lemma III: Let A and B be two polyhedra and let α_i , $i = 1,2,\ldots,r$, and β_j , $j = 1,2,\ldots,s$, denote their dihedral angles.

If $A \sim B$, then there exist positive integers m_i , n_j , $i = 1,2,\ldots,r$, $j = 1,2,\ldots,s$, and an integer k such that

(5.1) $$\sum_{i=1}^{r} m_i\alpha_i = \sum_{j=1}^{s} n_j\beta_j + k\pi .$$

Equation (5.1) is of the same form as (4.1); since in section 4 we have demonstrated that the dihedral angles of the polyhedra X and Y defined there cannot satisfy an equation of this form, it follows from Lemma III that not only $X \approx Y$ does not hold, but $X \sim Y$ does not hold either, which proves Theorem III.

The crux of the matter then is to prove Lemma III. This cannot be done in the simple and straightforward manner in which Lemma II was proved because there the regularity of the subdivisions was essentially used.

Preparatory remarks: We could attempt to show that $A \sim B$ implies $A \approx B$ and thus reduce Lemma III to Lemma II by obtaining from the irregular subdivisions $A = \Sigma A_i$, $B = \Sigma B_i$ regular ones. This can be attempted by introducing on A_i as new edges and vertices all incidences of edges and vertices of the other A_j with A_i in the sum ΣA_i ; let us denote the polyhedron thus obtained from A_i by A_i' ; then we have to introduce corresponding new edges and vertices on B_i obtaining B_i'.

Since the subdivision $B = \Sigma B_i'$ is in general not a regular one, we have to introduce all incidences of edges and vertices of B_j' with B_i' as new edges and vertices on B_i', obtaining the new polyhedra B_i''. This subdivision has to be transferred to the A_i'-s, and so on. Only if this process terminates in a finite number of steps will be obtain a common regular subdivision of A and B. In the following (for sake of simplicity two-dimensional) example the process terminates after the second step:

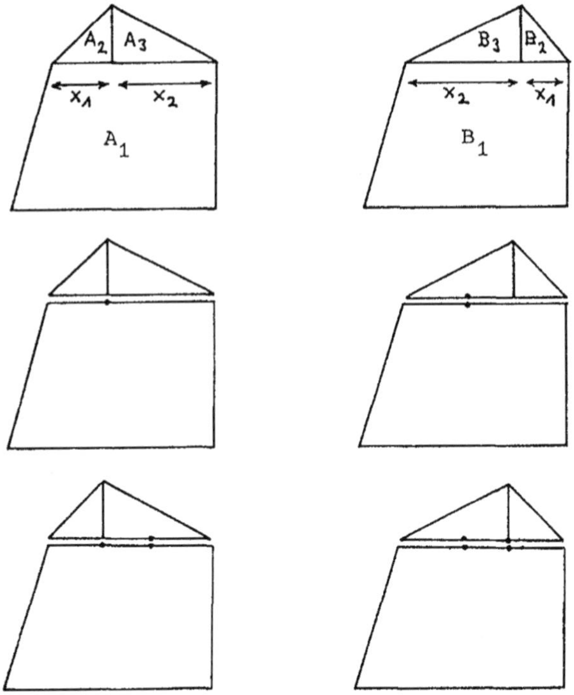

A B

A' B'

A" B"

We will give a proof of Lemma III which works with a common subdivision only on the <u>edges</u> of the polyhedra A_i and B_i.

<u>Proof of Lemma III</u>: (*) Assume that $A \sim B$. Then there exist polyhedra C_i, $i = 1,2,\ldots,n$, such that

$$A = \sum_{i=1}^{n} A_i \ , \quad B = \sum_{i=1}^{n} B_i \ , \quad A_i \cong B_i \cong C_i \ , \qquad i = 1,2,\ldots,n \ .$$

(*) This proof is due to Kagan, Math. Ann., Bd. 57.

In the sum $\sum_{i=1}^{n} A_i$ we consider all incidences of vertices of A_j with edges of A_i , i, j = 1,2,...,n , and denote the corresponding points on the congruent polyhedra C_i by a_j . We define the points b_j in a similar manner. These points a_j , b_j divide the edges of C_i into intervals which will be denoted by e_k . The points a_j also divide the edges of C_i into intervals e_k^a , and each e_k^a is the sum of a finite number of e_k . We define the intervals e_k^b in a similar manner. We assign to each e_k a positive integer p_k ; having done

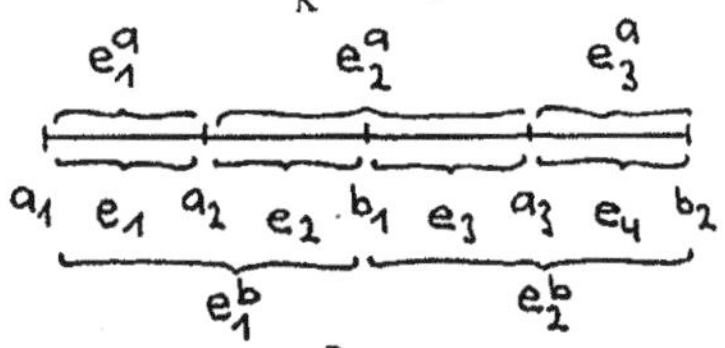

this we can assign to each $e_j^a = e_{r_1} + e_{r_2} + \ldots + e_{r_m}$ the positive integer $p_j^a = p_{r_1} + p_{r_2} + \ldots + p_{r_m}$ and similarly to each $e_j^b = e_{s_1} + e_{s_2} + \ldots + e_{s_n}$ positive integer $p_j^b = p_{s_1} + p_{s_2} + \ldots + p_{s_n}$.

Lemma IV: It is possible to assign a positive integer p_k to each e_k so that whenever $e_{k_1}^a$ and $e_{k_2}^a$ coincide in the sum ΣA_i

(5.2) $$p_{k_1}^a = p_{k_2}^a$$

holds, and similarly whenever $e_{k_1}^b$ and $e_{k_2}^b$ coincide in the sum ΣB_1

(5.3) $$p_{k_1}^b = p_{k_2}^b$$

holds.

Proof: We shall prove Lemma IV by making use of the following algebraic lemma:

Lemma V: If a system of homogeneous equations

(5.4) $$\sum_{j=1}^{n} c_{ij} x_j = 0 \quad , \quad i = 1,2,\ldots,m$$

with integer coefficients c_{ij} has a solution $X = (x_1, x_2, \ldots, x_n)$ for which $x_i > 0$ holds for $i = 1,2,\ldots,n$, then there exists another solution $Q = (q_1, q_2, \ldots, q_n)$ such that q_i is a positive integer, $i = 1,2,\ldots,n$.

Proof of Lemma V: According to the established theory of systems of linear equations all solutions of (5.4) can be written as the linear combination of a finite number of rational vectors X_h , h = 1,...,H, $H \leq n$. Consequently the solution $X = (x_1, x_2, \ldots, x_n)$ can be written in the form

$$X = (x_1,x_2,\dots,x_n) = \lambda_1 X_1 + \dots + \lambda_H X_H ;$$

consider

$$(x'_1,x'_2,\dots,x'_n) = \lambda'_1 X_1 +\dots+ \lambda'_H X_H$$

where the λ'_h are rational numbers. Since we assumed $x_i > 0$ for $i = 1,2,\dots,n$ we can choose $|\lambda_h - \lambda'_h|$, $h = 1,2,\dots,H$, so small that $x'_i > 0$ for $i = 1,2,\dots,n$. Since the vectors X_h are rational vectors and λ'_h were chosen to be rational, the numbers x'_i will also be rational. Since a constant multiple of a set of solutions of homogeneous equations is also a solution, we can multiply the x'_i by their least common denominator and obtain a set of positive integer solutions.

Q.e.d.

We observe that equations (5.2) and (5.3) are linear and homogeneous with integer coefficients in the p_k ; if we write x_k for p_k in these equations we can immediately verify that x_k = length of e_k is a solution of them. Since these x_k are > 0 , it follows from Lemma V that (5.2) and (5.3) possess positive integer solutions. This completes the proof of Lemma IV if we choose these values as our p_k .

To each e_k, e^a_k, and e^b_k we assign an angle φ_k, φ^a_k, and φ^b_k = the dihedral angle of that edge of C_i on which e_k, e^a_k, or e^b_k lies.

It follows from the definition that if e_{k_1} and e_{k_2} are parts of $e^a_{k_3}$, then

$$\varphi_{k_1} = \varphi_{k_2} = \varphi^a_{k_3}$$

and similarly if e_{ℓ_1} and e_{ℓ_2} are parts of $e^b_{\ell_3}$, then

$$\varphi_{\ell_1} = \varphi_{\ell_2} = \varphi^b_{\ell_3} .$$

Consider the sum

$$\Sigma\, p_k \varphi_k \tag{5.5}$$

where the summation is extended over all k . We divide the set of intervals e_k into <u>subsets</u> by grouping together all e_k that are part of the same e^a_ℓ . We group together all terms $p_k\varphi_k$ in the sum (5.5) that are associated with intervals e_k belonging to the same subset. This rearrangement shows that (5.5) can be written in the form

$$\Sigma\, \varphi^a_k p^a_k . \tag{5.6}$$

We shall evaluate (5.6) by grouping together all terms $\varphi^a_{k_i} p^a_{k_i}$

which are associated with intervals $e^a_{k_i}$ that are coincident in the sum ΣA_i. The positive integers $p^a_{k_i}$ are, by Lemma III, equal for all these intervals, therefore the sum $\Sigma\varphi^a_{k_i}p^a_{k_i}$ around each edge is equal to:

(a) $\pi p^a_{k_i}$ if the intervals $e^a_{k_i}$ lie on a face of one of the A_i.

(b) $2\pi p^a_{k_i}$ if $e^a_{k_i}$ lie on an interior edge in the sum ΣA_i which does not lie on a face.

(c) $p^a_{k_i}\alpha_i$ if $e^a_{k_i}$ lie on an edge of A the dihedral angle of which is α_i.

Therefore we see that the sum (5.6) is equal to

(5.7) $$\sum_i m_i\alpha_i + k_1\pi \ ,$$

the summation extending over all dihedral angles of A, where m_i denote positive integers and k_1 an integer.

Repeating this reasoning with the polyhedron B instead of A we get for the value of the sum (5.5)

(5.8) $$\Sigma\, n_j\beta_j + k_2\pi \ ,$$

the summation extending over all dihedral angles of B, where n_j denote positive integers, k_2 an integer. (5.1) follows from equating (5.7) and (5.8) and setting $k = k_2 - k_1$.

Q.e.d.

Corollary to Theorem III: We divide the set of all polyhedra of volume one into equivalence classes; the number of classes is at least 2. Dehn and recently Sydler (*) obtained a sharper result:

Theorem IV: The power of the class of equivalent polyhedra of unit volume is that of the continuum.

No proof of this theorem will be given here.

Section 6.

In the last paragraph we have shown that not all polyhedra of the same volume are equivalent; this brings up the following important and interesting problem: Characterize all polyhedra that belong to the same equivalence class.

(*) Sydler, J.P.: Sur la décomposition des polyèdres. Comment. Math. Helv. 16, 266-273 (1944).

This is a difficult problem and it has not been completely solved yet. Partial results in the form of necessary conditions for equivalence which are stronger than Lemma III have been obtained by Dehn.

We shall prove

<u>Theorem V</u>: All prisms of the same volume belong to the same equivalence class.

<u>Proof</u>: We divide the base of the prism P into polygons so that the diameter (i.e. the maximum distance of any two points) of each polygon is $< \frac{h}{2} \cos \alpha$, where h is the altitude, α the angle enclosed by the altitude and the generator.

We divide P into prisms P_i having as base the polygons into which the base of P was divided, and the same altitude and generator as P. We take any point of P_i at the altitude $h/2$ and construct a plane passing through this point and normal to the generator of P_i. This plane divides the prism into two parts P_i^1 and P_i^2.

Since the diameter of the base of P_i was assumed to be $< \frac{h}{2} \cos\alpha$, this plane will not intersect the base B_i^1 or the top B_i^2 of P_i. Therefore we put P_i^1 and P_i^2 together so that B_i^1 and B_i^2 coincide and obtain a right prism P_i' with altitude $h' = h/\cos\alpha$. It follows from the construction that $P_i \sim P_i'$. By Theorem I the base B_i' of P_i' is equivalent to a normal rectangle. Therefore, since P_i' is a right prism, P_i' – and consequently P_i – is equivalent to a right rectangular

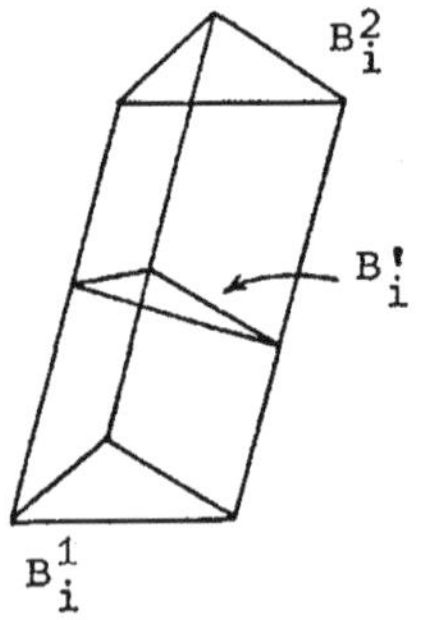

prism P_i'' two sides of which have the length 1 and h' units. Putting these prisms P_i'' together so that their 1 by h' face coincides we obtain a prism P'', two sides of which have the length 1 and h', such that $P \sim P''$. Using Theorem I once more we find a normal rectangular prism P''' such that $P \sim P'''$. Since the length of the third side of P''' is determined by the volume of P, we see that all prisms P

having the same volume are equivalent to the <u>same</u> prism P''' , hence by the transitivity of equivalence they belong to the same equivalence class.

Q.e.d.

<u>A generalized notion of equivalence</u>: Two polyhedra A and B are equivalent with respect to decomposition and completion, denoted by $A \ominus B$, if there exist two polyhedra C and C' , such that C and A, C' and B have no interior point in common and

$$C \sim C' \;,\; A + C \sim B + C' \;.$$

Obviously $A \sim B$ implies $A \ominus B$.

Sydler (*) has shown that $A \ominus B$ implies $A \sim B$, i.e. the notion of equivalence with respect to decomposition and completion is not more general than equivalence with respect to decomposition alone. The only advantage of the former is that $A \sim B$ can be demonstrated, if true, without the use of the axiom of Archimedes.

Two polyhedra A and B are <u>equivalent</u> and <u>equally oriented</u> if

$$A = \sum_{i=1}^{n} A_i \;,\quad B = \sum_{i=1}^{n} B_i \;,$$

A_i and A_j , B_i and B_j , $i \neq j$, having no interior point in common, and A_i is congruent and equally oriented to B_i for $i = 1,2,\ldots,n$.

Gerling has shown that two tetrahedra that are mirror images of each other can always be decomposed into a finite number of polyhedra that are pairwise congruent <u>and</u> equally oriented. Hence $A \sim B$ always implies that A and B are equivalent <u>and</u> equally oriented.

For a full discussion of equivalence and related topics we refer the reader to Sydler's paper.

(*) Sydler, J.P.: Sur la décomposition des polyèdres. Comment. Math. Helv., 16, 266-273 (1944).

PROBLEMS

Chapter I

(1) The polar G' of a corner G is defined as the corner whose edges are the positive normals to the faces of G. Derive Euler's relation for a convex polyhedron P by constructing the polars of the corners at all vertices of P and using the formulas for the area of plane and spherical polygons. (Hint: Find connection between face angles of G and dihedral angles of G'.)

(2) Generalize this proof for polyhedra of genus p.

(3) Generalize this proof to n-dimensional convex polyhedra.

(4) Construct on a special surface of genus p vector fields which have

(a) two singularities of index $+1$, and $2p$ singularities of index -1. (Hint: Consider the field as the gradient of a function with one maximum, one minimum and $2p$ saddlepoints)

(b) two singularities, each with index $1-p$.

(c) one singularity, with index $2-2p$.

(d) k singularities with prescribed indices $j_1, j_2, \ldots, j_k$, where $j_1 + j_2 + \ldots + j_k = 2-2p$.

(5) Consider the product

$$(1-f_{r_1}(x))(1-f_{r_2}(x)) \;\ldots\; (1-f_{r_s}(x))$$

where the $f_r(x)$ are the functions defined in section 14. Expand the product, integrate over the n-dimensional unit sphere and sum over all combinations of s functions $f_i(x)$. Express this result as a linear homogeneous relation between σ_r, $r = 0,1,\ldots,n$, where σ_r is the sum of the angles on all r dimensional faces of the simplex.

How many of these $n+2$ relations are linearly independent?

Chapter II

(1) Formulate and prove a theorem analogous to the four vertex theorems for polygons.

(2) Show by means of an explicit example that the four vertex theorem does not hold for self-intersecting curves. Consider the proof of the four vertex theorem given in Bieberbach's "Differentialgeo-

metrie" and show why it doesn't apply to self-intersecting curves.

Chapter IV

(1) (a) Show tetrahedron 1231' is not equivalent to a cube.

(b) Show tetrahedron 1233' is equivalent to a cube, and demonstrate this by decomposing the tetrahedron into no more than five parts.

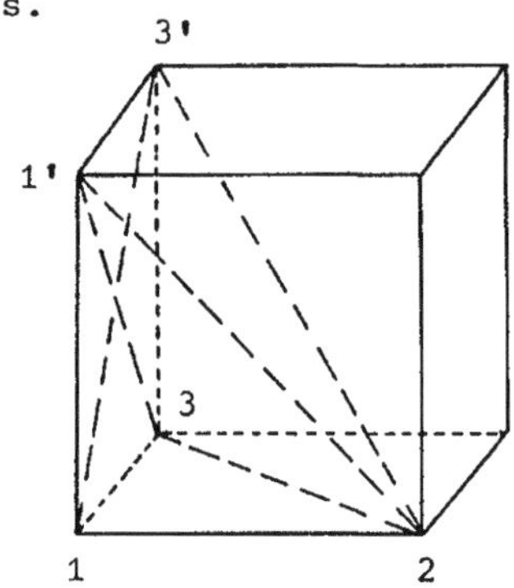

(2) A theorem of Euclid states that in the areas of rectangles I, I', and II, II' are equal. By theorem I we have

$$I \sim I' \quad , \quad II \sim II' \; .$$

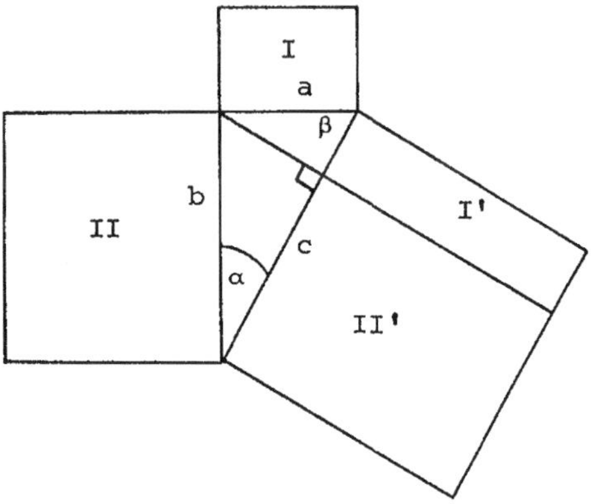

Demonstrate this directly by decomposing these regions into pairwise congruent polygons. What is the least number of parts, I and II respectively have to be decomposed into to demonstrate equivalence? (Express it in terms of the angles α, β) .

(3) Prove that two tetrahedra which are mirror images of each other are equivalent and equally oriented.

PART TWO

Differential Geometry in the Large

Stanford University 1956
Notes by J.W. Gray

CONTENTS

INTRODUCTION

This series of lectures will deal exclusively with the global geometry of two-dimensional surfaces. The intrinsic Riemannian geometry of surfaces will be considered only to a small extent, while the major concern will be with surfaces, especially closed ones, in three dimensional Euclidean space.

The first chapter will be a review of classical differential geometry in the small, and the second will be devoted to some general remarks on closed, differentiable surfaces, not necessarily in E^3. These will be followed by a short chapter on the Riemannian geometry of closed surfaces in which will be considered the relation between the Gauss curvature of a surface, the singularities of fields of directions on the surface and the topological structure of the surface. The remainder of the lectures will deal with surfaces in E^3.

The material covered in the first chapter can be found in greater detail in the following texts:

Struik, D.J., Classical Differential Geometry
Darboux, G., Leçons sur la théorie générale des surfaces
Blaschke, W., Vorlesungen über Differentialgeometrie, Vol. I.

CHAPTER I

Differential Geometry of Surfaces in the Small [summary sketch]

0. Notation

The following notation will be used:
There will be a "parameter plane", E^2 , with Cartesian coordinates (u,v) or (u^1,u^2) and a Euclidean space E^3 with Cartesian coordinates (x,y,z) or (x^1,x^2,x^3) .

Vectors (usually in E^3) will be denoted by Roman Capital letters, X, Y, etc. However, such letters will be used also for things which are not vectors. The meaning should be clear from the context. Juxtaposition of vectors denotes scalar product and $\times$ denotes vector product.

If $u = u(t)$, where t is a parameter, then $u' = \frac{du}{dt}$. If $x = x(u,v)$, then $x_u = \frac{\partial x}{\partial u}$ and $x_v = \frac{\partial x}{\partial v}$ and similarly if $x = x(u^1,u^2)$ then $x_1 = \frac{\partial x}{\partial u^1}$ and $x_2 = \frac{\partial x}{\partial u^2}$. If, further, $u = u(t)$ and $v = v(t)$, then $x' = x_u u' + x_v v'$.

If $X = (x,y,z)$ and $x,y,$ and z are functions of u and v , then $X_u = (x_u,y_u,z_u)$ and $X_v = (x_v,y_v,z_v)$. The definitions of X_1 , and X_2 are similar to those given above. If $X(t) = (x(t),y(t),z(t))$ is a curve, then $X' = (x',y',z')$. If $t = s$ is the arc length, then X' is denoted by $\dot{X}$ or $X^{\cdot}$.

Warning. Existence and continuity of derivatives will be assumed (without explicit mention) wherever this will expedite the statements and proofs of results. No special attempt will be made to always give the "best" results with respect to minimum conditions on derivatives. However, whenever analyticity is necessary, this will be stated explicitly. In general, most expressions which will be considered will require that the functions appearing in them be twice continuously differentiable to be meaningful and three times continuously differentiable to be provable.

1. Elementary Concepts

1.1 Definition. A surface in E^3 is a map of a region R in the u-v-plane (called the parameter plane) into E^3 subject to the following conditions: If the mapping is given by specifying the functions $x(u,v)$, $y(u,v)$, and $z(u,v)$ then

1) all first partial derivatives exist and are continuous (stronger assumptions will be made when necessary)

2)
$$\operatorname{rank}\begin{pmatrix} x_u & y_u & z_u \\ x_v & y_v & z_v \end{pmatrix} = 2 \quad .$$

The map will be denoted by $X(u,v) = (x(u,v), y(u,v), z(u,v))$.

1.2 Theorem. A map of a region R in the u-v-plane into E^3 which defines a surface is a local homeomorphism; i.e., the map gives a homeomorphism between a neighborhood of each point and the image of the neighborhood under the map.

Proof: We may assume
$$\begin{vmatrix} x_u & y_u \\ x_v & y_v \end{vmatrix} \neq 0 \quad .$$

But then the projection of the surface into the x-y-plane is a local homeomorphism, of the u-v-plane into the x-y-plane, since the Jacobian of this map is not zero. Therefore the map into the surface is locally 1-1 and open, and hence a local homeomorphism.

1.3 Definition. If $X = (x(u,v), y(u,v), z(u,v))$ is a surface, then X_u and X_v are tangent vectors to the surface. The plane spanned by X_u and X_v is called the tangent plane.

The vector $\bar{X} = \dfrac{X_u \times X_v}{|X_u \times X_v|}$ is the normal to the surface.

Condition 2) of 1.1 can be stated $X_u \times X_v \neq 0$; i.e. X_u and X_v do not vanish and have different directions.

1.4 Definition. A motion is a translation, a rotation, a reflection, or any combination of these three.

An admitted parameter transformation is a one-to-one map of a region $\bar{R}$ in the $\bar{u}$-$\bar{v}$-plane into a region R in the u-v-plane such that $\dfrac{\partial(u,v)}{\partial(\bar{u},\bar{v})} \neq 0$.

Under such a transformation, it is often necessary to decrease the region in which certain expressions are valid. This, in a sense, is characteristic of classical differential geometry, and is precisely the sort of argument which is not allowed in differential geometry in the large.

If $X(u,v)$ is a surface, then clearly under an admitted parameter transformation, $X[u(\bar{u},\bar{v}),v(\bar{u},\bar{v})]$ is also a surface whose image is the same point-set as the image of $X(u,v)$. If $\frac{\partial(u,v)}{\partial(\bar{u},\bar{v})} > 0$, then $X_{\bar{u}} \times X_{\bar{v}}$ and $X_u \times X_v$ have the same direction; otherwise, they have opposite directions.

2. First Fundamental Form

2.1 Definition. If $X(u,v)$ is a surface, then the expression

$$X_u^2 du^2 + 2X_u X_v du\, dv + X_v^2 dv^2$$

is called the first fundamental form of the surface. If $E = X_u^2$, $F = X_u X_v$, and $G = X_v^2$, then it can be written

$$E\, du^2 + 2F\, du\, dv + G\, dv^2 .$$

If coordinates (u^1,u^2) are used instead of (u,v), then the first fundamental form will be written

$$g_{ij} du^i du^j \quad \text{where} \quad g_{11} = E,\ g_{12} = g_{21} = F,\ g_{22} = G .$$

2.2. Let $u(t)$, $v(t)$ define a curve in the u-v-plane. Then, if $X(u,v)$ is a surface, $X(u(t),v(t))$ is a curve in the surface. The length of this curve is defined to be $\int\sqrt{(X')^2}dt$. Since $X' = X_u u' + X_v v'$ we have immediately that

$$\int\sqrt{(X')^2}dt = \int\sqrt{E(u')^2 + 2F\,u'v' + G(v')^2}dt$$
$$= \int\sqrt{E\,du^2 + 2F\,dudv + G\,dv^2} .$$

Hence, in the surface, the element of arc length, ds, is given by

$$ds^2 = E\,du^2 + 2F\,dudv + G\,dv^2 .$$

It follows from this that the first fundamental form is positive definite.

2.3. If two intersecting curves are given, then the angle between the curves is defined to be the angle between the tangents to the curves. Using coordinates (u^1,u^2), their tangent vectors can be expressed as linear combinations, $a^i X_i$ and $b^j X_j$ respectively, of the tangent vectors X_1 and X_2. If γ is the angle between the tangents to the curves, then

$$\cos\gamma = \frac{g_{ij}a^i b^j}{\sqrt{g_{ij}a^i a^j}\sqrt{g_{ij}b^i b^j}}$$

2.4. The area A of a region on a surface X is given by

$$A = \iint \sqrt{EG-F^2}\,dudv = \iint \sqrt{\det(g_{ij})}\,du^1du^2 \quad .$$

2.5. In the preceding discussion, it was shown that the first fundamental form of a surface essentially determines the metric properties of that surface. It is easily seen that these notions are invariant with respect to motions and admitted parameter transformations. In fact, if g_{ij} are components of the first fundamental form with respect to (u^1,u^2) and $\bar{g}_{hk}$ are the components with respect to $(\bar{u}^1,\bar{u}^2)$ then

$$g_{ij}du^idu^j = \bar{g}_{hk}d\bar{u}^hd\bar{u}^k$$

i.e., the g_{ij} transform like a second order, covariant symmetric tensor. If X and Y are two surfaces which can be transformed into each other by a motion, then they have the same first fundamental forms, but the converse is not true.

2.6 Definition. Let X and Y be two surfaces. If there is a differentiable homeomorphism between X and Y, which preserves the length of curves, then X and Y are said to be isometric. The map is called an isometry.

If parameters u and v are introduced so that X(u,v) and Y(u,v) are corresponding points under a given homeomorphism h, then h is an isometry if and only if the first fundamental forms of X and Y with respect to u and v are the same.

3. Geodesic Lines

3.1 Definition. Let X be a surface and let X(s) be a curve on the surface where s is arc length. If $\ddot{X} = \lambda \bar{X}$ then X(s) is called a geodesic. If u and v are the surface parameters, then this condition is equivalent to $\ddot{X}X_u = 0$ and $\ddot{X}X_v = 0$. By eliminating arc length between these two equations and taking v as the parameter along the curve, it can be shown that the curve satisfies a second order differential equation

$$u'' = f(u',u,v) \quad .$$

Thus, the following theorem follows trivially.

3.2 Theorem. Given a point on a surface and a direction in the surface, there exists exactly one geodesic line through the point in the given direction.

<u>3.3 Theorem</u>. A curve $X(s)$ is a geodesic if and only if for each pair of points $\underline{a}$ and $\underline{b}$ on $X(s)$, the length ℓ of $X(s)$ between $\underline{a}$ and $\underline{b}$ is stationary with respect to variation of the curve between $\underline{a}$ and $\underline{b}$.

<u>Proof</u>: Let $X(s,\tau)$ be a family of curves such that $X(s,0) = X(s)$, and let

$$\ell(\tau) = \int_a^b \sqrt{[\dot{X}(s,\tau)]^2}\,ds .$$

We wish to show that $\ell'(0) = 0$.
A simple computation shows

$$\ell'(0) = \int_a^b \frac{\dot{X}\dot{X}'ds}{\sqrt{(\dot{X})^2}} .$$

But $(\dot{X})^2 = 1$ since s is arc length. Since $(\dot{X}X')^{\dot{}} = \ddot{X}X' + \dot{X}\dot{X}'$, by integration by parts, we get

$$\ell'(0) = [\dot{X}X']_a^b - \int_a^b \ddot{X}X'ds .$$

Now $\dot{X}$ is the tangent to the curve $X(s)$ and X' is the vector in the direction of the variation. Since we are keeping the end points, $\underline{a}$ and $\underline{b}$, fixed, $[\dot{X}X']_a^b = 0$. Further, since X' is a tangent vector, if Y is a tangent vector orthogonal to $\dot{X}$, we may write $X' = p(s)\dot{X} + q(s)Y$. Finally, since $\dot{X}^2 = 1$, $\ddot{X}\dot{X} = 0$. Hence

$$\ell'(0) = -\int_a^b (\ddot{X}Y)q(s)\,ds .$$

If $\ell'(0)$ is to be 0 for every function $q(s)$, which vanishes at $\underline{a}$ and $\underline{b}$, then by the Fundamental Principle of the Calculus of Variation, we must have $\ddot{X}Y = 0$. But, since trivially $\ddot{X}\dot{X} = 0$, this is equivalent to the condition $\ddot{X} = \lambda\bar{X}$.

On the other hand, if $\ddot{X} = \lambda\bar{X}$, then $\ddot{X}Y = \lambda\bar{X}Y = 0$, and hence $\ell'(0) = 0$.

<u>3.4 Definition</u>. Let C be an arbitrary curve on a surface X. Let v be arc length along C. By 3.2 through each point of C, there is a unique geodesic orthogonal to C. By well-known properties of differential equations, these geodesics depend continuous on their point of intersection with C. Hence if u measures arc length along these geodesics, then u and v give an admitted parameter system in a neighborhood of C where the geodesics do not intersect. This coordinate system is called a <u>geodesic coordinate system</u>, and u and v

are called <u>geodesic parameters.</u>

<u>3.5</u> <u>Theorem</u>. If u and v are geodesic parameters on X, then $ds^2 = du^2 + g^2dv^2$. Conversely, if $ds^2 = du^2 + g^2dv^2$, then the curves v = constant are geodesics.

<u>Proof</u>: In general $ds^2 = E\,du^2 + 2F\,dudv + G\,dv^2$. If u and v are geodesic parameters, then $E = X_u^2 = 1$ since u is arc length. We must show that $F(u,v) = X_uX_v = 0$. Now $F(0,v) = 0$ by definition. It is sufficient to show that $F_u = 0$. But

$$F_u = X_{uu}X_v + X_uX_{uv} .$$

However, $X_uX_{uv} = \frac{1}{2}E_v = 0$, since $E = 1$, and $X_{uu} = \ddot{X}$ which is normal to the surface since the u-curves are geodesics. Since X_v is a tangent vector, $X_{uu}X_v = 0$. Therefore $F_u = 0$. Finally if $g = +\sqrt{G}$, then

$$ds^2 = du^2 + g^2dv^2 .$$

Conversely, let $\underline{a}$ and $\underline{b}$ be two points on a curve v = constant and let D be any curve joining them. Then

$$\ell(D) = \int_a^b \sqrt{du^2 + g^2dv^2} \geq \int du = \ell(U)$$

where U is the curve v = constant. Therefore U is a curve of minimum length between $\underline{a}$ and $\underline{b}$, and hence a geodesic.

<u>4. Parallel Displacement</u>

<u>4.1</u> <u>Definition</u>. Let s be arc length along a curve C in a surface X. Let $Z(s)$ be a tangent vector field along C. Then Z is called a <u>parallel vector field</u> if the tangential component of $\dot{Z}$ is zero; i.e., $\dot{Z}_{tang} = 0$. More precisely, this is parallelism in the sense of Levi-Civita.

<u>Example</u>: If C is a geodesic, then $\ddot{X}_{tang} = 0$ since $\ddot{X}$ is normal to the surface. Hence the tangent vectors, $\dot{X}$, to a geodesic form a parallel vector field.

<u>4.2</u> <u>Theorem</u>. If Z_1 and Z_2 are two parallel vector fields, then Z_1Z_2 = constant.

<u>Proof</u>: $(Z_1Z_2)^{\cdot} = \dot{Z}_1Z_2 + Z_1\dot{Z}_2 = 0$ since $\dot{Z}_1$ and $\dot{Z}_2$ are normal, while Z_1 and Z_2 are tangential.

Since Z^2 = constant is a special case of the theorem, the length of a parallel vector field is constant. Also, the angle between two parallel vector fields is constant. Hence it is meaningful to speak of the parallel displacement of the whole vector bundle at a point, since this displacement is rigid.

__4.3__ __Theorem__. Given a curve C on a surface $X(u^1,u^2)$ and a vector at a point of C, there exists exactly one parallel vector field along C containing the given vector.

__Proof__: If $Z(s)$ is a parallel vector field along C, we wish to show that the condition $\dot{Z}_{tang} = 0$ is equivalent to a system of first order differential equations. Clearly $\dot{Z}_{tang} = 0$ is equivalent to $\dot{Z}X_k = 0$, $k = 1, 2$. Since Z is a tangent vector,

$$Z = z^i X_i$$

where z^i are the components of Z. Hence

$$\dot{Z} = \dot{z}^i X_i + z^i \dot{X}_i$$
$$= \dot{z}^i X_i + X_{ij}\dot{u}^j z^i \quad .$$

Therefore, we must have

$$0 = \dot{z}^i X_i X_k + (X_{ij}X_k)\dot{u}^j z^i \; .$$

Now $X_i X_k = g_{ik}$. Thus if (g^{hj}) denotes the inverse matrix of (g_{ik}), multiplying by g^{hk} gives

$$0 = \dot{z}^h + g^{hk}X_{ij}X_k\dot{u}^j z^i \; .$$

If we let $\Gamma^h_{ij} = g^{hk}X_{ij}X_k$, then

$$0 = \dot{z}^h + \Gamma^h_{ij}\dot{u}^j z^i$$

which is the desired system of first order differential equation. Hence existence and uniqueness follow by well-known properties of such systems.

__4.4__ __Theorem__. In the special case that (u,v) is a geodesic coordinate system, i.e., $ds^2 = du^2 + g^2 dv^2$, then the differential equations reduce to

$$\dot{\alpha} = -g_u \dot{v}$$

where α is the angle between X_u and Z.

Proof: For a geodesic coordinate system we may write

$$Z = \cos\alpha\, X_u + \sin\alpha\, g^{-1}X_v$$

since $X_u^2 = 1$ and $X_v^2 = g^2$ and since we know from 4.2 that $|Z| = 1$. Then

$$\dot{Z} = \cos\alpha(\dot{X}_u + \dot{\alpha}\, g^{-1}X_v) + \sin\alpha(-\dot{\alpha}\, X_u + (g^{-1}X_v)^{\cdot}) .$$

But $X_u\dot{X}_u = 0$ and $g^{-1}X_v(g^{-1}X_v)^{\cdot} = 0$ since $X_u^2 = 1$ and $(g^{-1}X_v)^2 = 1$; and $X_uX_v = 0$, since (u,v) is a geodesic coordinate system. Hence

$$\dot{Z}\, X_u = \sin\alpha\,(-\dot{\alpha} + (g^{-1}X_v)^{\cdot}X_u) = 0$$

$$\dot{Z}\, X_v g^{-1} = \cos\alpha(\dot{\alpha} + \dot{X}_u(g^{-1}X_v)) = 0 .$$

However $\dot{X}_u g^{-1}X_v + X_u(g^{-1}X_v)^{\cdot} = (X_u g^{-1}X_v)^{\cdot} = 0$ since $X_u g^{-1}X_v = 0$. Hence the coefficients of $\sin\alpha$ and $\cos\alpha$ are the same. But $\sin\alpha$ and $\cos\alpha$ are never simultaneously zero. Therefore we conclude that

$$\dot{\alpha} + \dot{X}_u(g^{-1}X_v) = 0 .$$

Or, in terms of u and v ,

$$\dot{\alpha} + g^{-1}X_{uu}X_v\dot{u} + g^{-1}X_{uv}X_v\dot{v} = 0 .$$

Now $X_{uu}X_v + X_uX_{uv} = (X_uX_v)_u = 0$. But, since $X_uX_{uv} = 0$ also, it follows that $X_{uu}X_v = 0$. Finally, since $X_v^2 = g^2$, $X_{uv}X_v = \frac{1}{2}(g^2)_u = gg_u$. Therefore

$$\dot{\alpha} = -g_u\dot{v} .$$

We see from this result, that for a geodesic coordinate system, the equation for parallel displacement depends only on the first fundamental form. Since parallel displacement does not depend on the coordinate system, this is true in any coordinate system. However, we shall give a formal proof of this fact, in order to be able to use the results in Section 5.

4.5 Theorem. The equations for parallel displacement of a vector depend only on the first fundamental form.

Proof: It is sufficient to show that Γ_{ij}^h is determined by the first fundamental form. Since $\Gamma_{ij}^h = g^{hk}X_{ij}X_k$, it suffices to consider the quantities

$$\Gamma_{ij,k} = X_{ij}X_k .$$

Since $g_{ik} = X_iX_k$, the $\Gamma_{ij,k}$ satisfy the following properties,

1) $\Gamma_{ij,k} = \Gamma_{ji,k}$

2a) $\Gamma_{ij,k} + \Gamma_{kj,i} = \dfrac{\partial g_{ik}}{\partial u^j}$

b) $-\Gamma_{jk,i} - \Gamma_{ik,j} = -\dfrac{\partial g_{ji}}{\partial u^k}$

c) $\Gamma_{ki,j} + \Gamma_{ji,k} = \dfrac{\partial g_{kj}}{\partial u^i}$.

Adding equation 2a, b, and c, making use of 1) gives

$$\Gamma_{ij,k} = \frac{1}{2}\left(\frac{\partial g_{ik}}{\partial u^j} + \frac{\partial g_{kj}}{\partial u^i} - \frac{\partial g_{ji}}{\partial u^k}\right)$$

5. Riemannian Space

5.1 Definition. Let $ds^2 = g_{ij}du^i du^j$ be a Riemannian metric in a region of the u-v-plane. (i.e., the g_{ij} form a positive definite quadratic form). Then as in Section 4, we define

$$\Gamma_{ij,k} = \frac{1}{2}\left(\frac{\partial g_{ik}}{\partial u^j} + \frac{\partial g_{kj}}{\partial u^i} - \frac{\partial g_{ij}}{\partial u^k}\right)$$

$$\Gamma^h_{ij} = g^{hk}\,\Gamma_{ij,k} .$$

If z^i are the components of a field of contravariant vectors on a curve which satisfies the equations

$$\dot{z}^h + \Gamma^h_{ij}\dot{u}^j z^i = 0$$

then the field is called a parallel field. The equations are called the equations of parallel displacement.

A curve is called a geodesic if the tangent vectors to the curve satisfy the equation of parallel displacement.

5.2 Theorem. Let A and B be vectors with components a^i and b^i respectively. Then the scalar product $AB = g_{ij}a^i b^j$ is invariant under parallel displacement. This property is characteristic in the sense that if the $\Gamma_{ij,k}$ are allowed to be arbitrary functions of u and v which are symmetric in the first two indices and which preserve scalar products, then they must be the $\Gamma_{ij,k}$ given in 5.1.

Exercise: Prove Theorem 5.2.

5.3 Theorem. Geodesics satisfy a second order differential equation and the distance along a geodesic between two of its points is stationary with respect to variations of the curve between the two points. As in Section 3, geodesic coordinates can be introduced along a curve.

6. Curvature in Two Dimensional Riemannian Geometry

6.1. Let the closed curve C be the boundary of a region R on a Riemannian manifold, U a continuous field of directions on $R \cup C$, Z_0 a vector ($\neq 0$) in a point of C, Z_t $(0 \leq t \leq 1)$ the field generated by parallel displacement of Z_0 around C, $\alpha = \measuredangle[U, Z_t]$. Then the variation $\delta_C \alpha$ along C equals mod. 2π the angle $\measuredangle[Z_0, Z_1]$. It is easily shown that $\delta_C \alpha$ does neither depend on the choice of U nor on the choice of Z_0.

6.2 Theorem. Let R be a region small enough to be contained in the region of validity of a geodesic coordinate system and let C be the boundary of R. Then

$$\delta_C \alpha = -\iint_R \frac{g_{uu}}{g} \, dA$$

where dA is the surface element, and hence $dA = g\,du\,dv$.

Proof: By applying 4.4, we get $\delta_C \alpha = \oint_C \dot{\alpha} ds = -\oint_C g_u \dot{v} \, ds = -\oint_C g_u dv$

$$= -\iint_R g_{uu} du \, dv = -\iint_R \frac{g_{uu}}{g} \, dA \quad .$$

6.3 Definition. $K(a) = -\dfrac{g_{uu}(a)}{g(a)}$ is called the Gauss-Riemann curvature. $\iint_R K \, dA$ is called the total curvature of the region R. By 6.2, the total curvature of a region R is equal the change in angle of a vector displaced parallely around the boundary of R.

6.4 Theorem. $K(a)$ is independent of the coordinate system at a.

Proof: $K(a) = -\dfrac{g_{uu}(a)}{g(a)} = \lim\limits_{C \to a} \dfrac{\delta_C \alpha}{A}$ where A is the area bounded by C and $\lim\limits_{C \to a}$ means limit as the curve C is shrunk to the point a. Since A and $\delta_C \alpha$ are independent of the coordinate system, the theorem follows.

6.5 Theorem. If K is constant, then, subject to initial value conditions, there is exactly one Riemannian metric with the given K. The cases $K>0$, $K<0$, and $K = 0$ are called respectively elliptical, hyperbolic, and Euclidean.

Proof: The proof follows easily since g satisfies the differential equation $g_{uu} + Kg = 0$.

6.6 Theorem. In an orthogonal coordinate system (i.e., $F = 0$), if $E = e^2$ and $G = g^2$, then

$$K = -\frac{1}{eg}\left[\left(\frac{e_v}{g}\right)_v + \left(\frac{g_u}{e}\right)_u\right]$$

$$= -\frac{1}{2\sqrt{EG}}\left[\left(\frac{E_v}{\sqrt{EG}}\right)_v + \left(\frac{G_u}{\sqrt{EG}}\right)_u\right]$$

7. The Gauss Curvature of Surfaces in E^3

7.1. In order to motivate the discussion of the curvature of a surface in E^3 , we shall review the definition of the curvature of a plane curve. If $C(s)$ is a plane curve, where s is the arc length and if τ is the angle between the tangent to the curve and the positive x-axis, then the curvature, k , is defined by $k = \dot{\tau} = \frac{d\tau}{ds}$. Clearly τ could as well be defined to be the angle between the normal to the curve and the positive x-axis. Let Γ be the unit circle in the x-y-plane and consider the mapping of $C(s)$ into Γ given by mapping $C(s_o)$ into the point on Γ which is the intersection of Γ with a unit vector through $(0,0)$ parallel to the normal to $C(s)$ at s_o . Let $\ell(s,s_o)$ be the length of $C(s)$ between s and s_o , and let $\bar{\ell}(s,s_o)$ be the length of the image of the arc between s and s_o under the map given above. Then

$$\bar{\ell}(s,s_o) = \int_{s_o}^{s} d\tau = \int_{s_o}^{s} k\, ds .$$

Hence

$$k(s_o) = \lim_{s\to s_o} \frac{\bar{\ell}(s,s_o)}{\ell(s,s_o)}$$

7.2. For a surface $X(u,v)$ in E^3 , we have a similar map into the uni sphere, Σ , defined by mapping the point $X(u_o,v_o)$ into the point on Σ which is the intersection of Σ with a unit vector through $(0,0,0)$ parallel to the normal to $X(u,v)$ at (u_o,v_o) . This map is called the spherical map of X . Let $A(u_o,v_o)$ be the area of a small region containing $X(u_o,v_o)$ and let Ω be the area of the image of this region under the spherical map. Then the curvature, K , is defined by

$$K(u_o,v_o) = \lim_{A\to o} \frac{\Omega}{A} = \frac{d\Omega}{dA} .$$

Gauss proved (Theorema Egregium) that this K depends only on the first fundamental form of the surface. It can be shown, in fact, that this K and the K defined in 6.3 are the same. From this, it follows immediately that if X and Y are surfaces which are isometric, then they have the same curvature at corresponding points.

8. The Second Fundamental Form

8.1 Definition. Let $X(u,v)$ be a surface. Then the second fundamental form of X is

$$X_{uu}\bar{X}\,du^2 + 2X_{uv}\bar{X}\,dudv + X_{vv}\bar{X}\,dv^2 .$$

Since $X_u\bar{X} = 0$ and therefore $X_{uu}\bar{X} = -X_u\bar{X}_u$, the second fundamental form can be written

$$-X_u\bar{X}_u du^2 - 2X_u\bar{X}_v dudv - X_v\bar{X}_v dv^2 .$$

If $L = -X_u\bar{X}_u$, $M = -X_u\bar{X}_v$, and $N = -X_v\bar{X}_v$ then it can also be written

$$L\,du^2 + 2M\,dudv + N\,dv^2 .$$

If coordinates (u^1,u^2) are used instead of (u,v) , then the second fundamental form is written

$$\ell_{ij}du^i du^j$$

where $\ell_{11} = L$, $\ell_{12} = \ell_{21} = M$, and $\ell_{22} = N$.

The second fundamental form will be used to derive detailed information about the curvature in the neighborhood of a point. We shall also be concerned with the entities $\ell_i^h = \ell_{ik}g^{kh}$, because of the following theorem.

8.2 Theorem. $\bar{X}_i = -\ell_i^j X_j$, i.e., the ℓ_i^j are the components of the linear transformation which gives the rate of change of the normal to the surface.

Proof: Since $\bar{X}_i$ is a tangent vector, $\bar{X}_i = a_i^j X_j$. Taking the scalar product with X_k gives $-\ell_{ik} = a_i^j g_{jk}$, and hence, multiplying by g^{kh} gives

$$a_i^h = -\ell_{ik}g^{kh} = -\ell_i^h$$

8.3. Since (ℓ_i^j) is a linear transformation we may consider its invariants under linear transformations; i.e., the determinant, the trace and the eigenvalues. It can be shown that

$$\det(\ell^j_i) = \frac{\det(\ell_{ij})}{\det(g_{ij})} = K ,$$

where K is the Gauss-Riemann curvature.

We define the mean curvature H by

$$H = \frac{1}{2}\,\mathrm{tr}(\ell^j_i)$$

where tr denotes trace. It can be shown easily that

$$2H = \frac{GL-2FM+EN}{EG-F^2}$$

If we choose parameters at a point so that $E = G = 1$ and $F = 0$ then $(g_{ij}) = (\delta_{ij})$ and $(g_{ij})^{-1} = (\delta_{ij})$. Then $(\ell^j_i) = (\ell_{ij})$ which is symmetric. Therefore (ℓ^j_i) has real eigenvalues, which will be denoted by k_1 and k_2 . k_1 and k_2 are called the two <u>principle curvatures</u> and satisfy the following equations:

$$k_1k_2 = K \quad\text{and}\quad k_1 + k_2 = 2H$$

and therefore $k_i = H \pm \sqrt{H^2-K}$. If the corresponding eigenvectors are uniquely defined (i.e., $k_1 \neq k_2$) , then the directions of the eigenvectors are called the <u>principle directions</u>. Points where $k_1 = k_2$ are called <u>umbilical points</u>. In the first case it is possible to find two families of curves such that the tangents to the curves are in the principle direction at each point. These curves are called the <u>lines of curvature</u>.

<u>8.4</u>. The equations satisfied by the lines of curvature can be derived as follows:

Let (u^1,u^2) be a coordinate system and (du^1,du^2) one of the principle directions. Then, since the direction is not changed by the spherical map, (du^1,du^2) must be proportional to $(\ell^1_j du^j,\ \ell^2_j du^j)$.

Hence

$$\begin{vmatrix} du^1 & du^2 \\ \ell^1_j du^j & \ell^2_j du^j \end{vmatrix} = 0 .$$

Expanding this in terms of (u,v) rather than (u^1,u^2) gives

$$\ell^2_1 du^2 + (\ell^2_2 - \ell^1_1)\,du\,dv - \ell^1_2 dv^2 = 0 .$$

Clearly, parameters can be introduced along these lines in the neighborhood of any point which is not an umbilical (or flat) point. It can be shown that a necessary and sufficient condition for the constant lines of a coordinate system to be lines of curvature is that $F = M = 0$.

8.5 Definition. Let (u,v) be a Euclidean orthogonal coordinate system in the tangent plane to the surface, with $(0,0)$ corresponding to the point on the surface. The conics given by

$$Lu^2 + 2M\,uv + Nv^2 = \pm 1$$

are called the Dupin indicatrix. There are several cases to consider:

1) The second fundamental form is definite; i.e. $LN - M^2 > 0$. Then only one choice of sign gives a real curve, which is an ellipse. In this case the tangent plane is entirely on one side of the surface. In a neighborhood of such a point, the spherical map is 1-1 and preserves orientation.

2) The second fundamental form is indefinite; i.e. $LN - M^2 < 0$. The locus is two conjugate hyperbolas. In this case the tangent plane intersects the surface. In a neighborhood of such a point the spherical map is 1-1 and reverses orientation

3) $LN - M^2 = 0$.

a) $(L,M,N) \neq (0,0,0)$. In this case the locus is two parallel straight lines since the left side of the equation is the square of a linear form.

b) $(L,M.N) = (0,0,0)$. There is no locus. Such a point is called a flat point.

No general statement can be made about the location of the tangent plane and the spherical map may not even be 1-1 .

8.6. The principle curvatures can be given a more geometric interpretation. For, given a point on the surface and a direction in the tangent plane to the surface at the point, a unique plane is determined by this direction and the normal to the surface at the point. The intersection of this plane with the surface is a plane curve whose curvature k can be shown to be

$$k = \frac{L\,du^2 + 2M\,dudv + N\,dv^2}{E\,du^2 + 2F\,dudv + G\,dv^2} .$$

k is a function of the direction in the tangent plane and it can be shown that $|k| = \frac{1}{r^2}$ where r is distance from the origin to the indicatrix. Hence, in general, k has one maximum and one minimum. These are precisely the eigenvalues k_1 and k_2, and the directions in which they occur coincide with the directions of the eigenvectors. Exceptions to this behavior occur when the point is an umbilical (or flat) point; then all directions are eigenvectors. In cases 2) and 3a) the asymptotic directions are given by

$$Lu^2 + 2M\,uv + Nv^2 = 0$$

8.7. The curvature K was introduced by means of the spherical map as a generalization of the curvature k of a plane curve. The mean curvature H is also a natural generalization of k, where this generalization is given by considering the variation of arc length of a curve between two points.

Let X be a curve in the plane between two points $\underline{a}$ and $\underline{b}$ of length ℓ. If X is varied, then, as in Section 3, the rate of change of ℓ is given by

$$\ell' = -\int_a^b \ddot{X}X'ds + [\dot{X}X']_a^b .$$

Let Y be the unit normal to the curve and let τ be the angle between the tangent to the curve and the positive x-axis. Then

$$\dot{X} = (\cos\tau, \sin\tau) .$$

So
$$\ddot{X} = k(-\sin\tau, \cos\tau) = kY$$

Since X' can be written

$$X' = t\dot{X} + nY$$

it follows that for an arbitrary variation X',

$$\ell' = -\int_a^b nk\,ds + [t]_a^b .$$

If X' is a normal variation of constant magnitude 1 ; i.e. $t = 0$ and $n = 1$, then

$$\ell' = -\int_a^b k\,ds .$$

If the analogous argument is carried out for a region of a surface X in E^3, then the result is in terms of H. Let p be the variation parameter and let $A(p)$ be the area of the varied surface. If X' is the derivative in the direction of the variation, then

$$X' = n\bar{X} + X'_{tang} .$$

It can be shown that

$$A' = -2\iint nH\, dA + \oint (\bar{X}, X', dX)$$

where (X,Y,Z) denotes the determinant of the vectors X, Y, and Z. If $X'_{tang} = 0$ and $n = 1$, then

$$A' = -2\iint H\, dA .$$

Exercise: Derive the above formula for A'.

9. The Relation between the two Fundamental Forms

9.1 Theorem. Let X and Y be two surfaces such that there is a 1-1 map of X onto Y. Let (u,v) be a coordinate system such that $X(u,v)$ and $Y(u,v)$ are corresponding points under the map. Then the first and second fundamental forms of X and Y with respect to u and v are the same, if and only if the map is produced by a proper motion. (Under a reflection, the first fundamental form is the same, while the second fundamental form changes sign).

9.2 Definition. Let $E\, du^2 + 2F\, dudv + G\, dv^2$ and $L\, du^2 + 2M\, dudv + N\, dv^2$ be two differential forms. Let Γ^k_{ij} be defined as in 5.1. Then the equations

$$L_v - M_u = \Gamma^1_{12}L + (\Gamma^2_{12} - \Gamma^1_{11})M - \Gamma^2_{11}N$$
$$M_v - N_u = \Gamma^1_{22}L + (\Gamma^2_{22} - \Gamma^1_{21})M - \Gamma^2_{21}N$$

are called the Codazzi equations.

9.3 Theorem. Given two forms, $E\, du^2 + 2F\, dudv + G\, dv^2$ and $L\, du^2 + 2M\, dudv + N\, dv^2$, there exists a surface with these as first and second fundamental forms respectively if and only if

1) $E\, du^2 + 2F\, dudv + G\, dv^2$ is positive definite
2) $LN - M^2 = f(E,F,G)$ where f is the operator given by Gauss' Theorema Egregium.
3) The forms satisfy the Codazzi equations (i.e., the Codazzi equations are essentially integrability conditions for the forms.)

Proof: We will sketch the proof of necessity. 1) and 2) are obvious. To prove 3) consider the following formulas:

$$X_{ij} = a^k_{ij}X_k + \ell_{ij}\bar{X} .$$

Taking the scalar product with X_h gives

$$\Gamma_{ij,h} = X_{ij}X_h = a_{ij}^k g_{kh} \quad .$$

Hence $a_{ij}^k = \Gamma_{ij}^k$. Therefore

$$X_{ij} = \Gamma_{ij}^k X_k + \ell_{ij}\bar{X} \quad .$$

We also know that $\bar{X}_i = -\ell_i^k X_k$. Now, assuming continuity of the third derivatives, $X_{ijk} = X_{ikj} = X_{kij}$, etc. By computation, one shows that these equations just reduce to the Codazzi equation and Gauss' theorema egregium.

10. Miscellaneous Remarks

10.1. We have seen that the coefficients of the first fundamental form g_{ij} are the components of a covariant symmetric tensor; i.e., if $g = (g_{ij})$ in one coordinate system and $\bar{g} = (\bar{g}_{ij})$ in another coordinate system, then $g = T'\bar{g}T$, where T is the Jacobian of the coordinate transformation and T' is the transpose of T .

However, the situation is not quite the same for the coefficients of the second fundamental form, ℓ_{ij} . Since $\ell_{ij} = X_{ij}\bar{X}$, the ℓ_{ij} would be the components of a covariant tensor if $\bar{X}$ was invariant under a change of coordinates. But, by our definition, $\bar{X}$ changes sign under a coordinate transformation which reverses orientation. Hence, in general, if $\ell = (\ell_{ij})$ in one coordinate system and $\bar{\ell} = (\bar{\ell}_{ij})$ in another coordinate system, then $\ell = \sigma_T T'\bar{\ell}T$ where σ_T is the sign of the determinant of T .

Now, the curvature K is not affected by this property, but the mean curvature H is, and in fact changes sign under an orientation reversing transformation, since $\mathrm{tr}(-\ell) = -\mathrm{tr}(\ell)$. Therefore the mean curvature of a surface is well-determined only if an orientation is chosen, or equivalently, only if the direction of the normal is specified.

10.2. In Section 3 it was shown that parameters could be introduced so that

$$ds^2 = du^2 + g^2 dv^2 \quad .$$

A second convenient parameter system which can be introduced is an isothermal parameter system, which satisfies $E = G$ and $F = 0$. If a coordinate system is given such that

$$ds^2 = E\,du^2 + 2F\,dudv + G\,dv^2$$

then a simple proof of the possibility of introducing isothermal parameters, u^* and v^* can be given providing E, F, and G are analytic, where the proof is given by continuing E, F, and G into the complex domain. The theorem is true under much weaker conditions - for example, if E, F and G are only twice continuously differentiable - but the proof is much more difficult.

Clearly such a mapping of the u^*- v^* - plane into a surface preserves angles and hence is conformal. In this case, when it is convenient we can introduce new parameters

$$w = u^* + iv^* \quad \text{and} \quad \bar{w} = u^* - iv^* .$$

10.3. In the special case that the surface is given by $z = z(x,y)$, the basic quantities can be explicitly calculated. Let p, q, r, s, and t denote respectively z_x, z_y, z_{xx}, z_{xy}, and z_{yy}. Then

$$E = 1 + p^2 \qquad F = pq \qquad G = 1 + q^2$$

$$L = \frac{r}{\sqrt{1+p^2+q^2}}, \qquad M = \frac{s}{\sqrt{1+p^2+q^2}}, \qquad N = \frac{t}{\sqrt{1+p^2+q^2}}$$

$$K = \frac{rt-s^2}{(1+p^2+q^2)^2}, \qquad 2H = \frac{(1+q^2)r-2pqs+(1+p^2)t}{(1+p^2+q^2)^{3/2}} .$$

CHAPTER II

Some General Remarks on Closed Surfaces in Differential Geometry

As a general reference to the topological material covered in this chapter, see Seifert-Threlfall: Lehrbuch der Topologie.

1. Simple Closed Surfaces in E^3

1.1 Definition. A *simple, closed* (i.e., compact) *surface* or *2-manifold in* E^3 is a set $S \subset E^3$ such that:

1) S is compact (i.e., closed and bounded)

2) S is connected. (A compact set S is not connected if $S = A \cup B$ where A and B are compact and non-empty, and $A \cap B$ is empty.)

3) Each point $p \in S$ has a neighborhood $N(p) \subset S$ which is homeomorphic to the interior of a disk in the plane.

S is called *differentiable* if in addition the following conditions are satisfied:

4) Let $N(p)$ be a neighborhood of p satisfying 3) above and let the homeomorphism be given by

$$X(u,v) = (x(u,v),\ y(u,v),\ z(u,v)),$$

where (u,v) are Euclidean coordinates in the plane and (x,y,z) are Euclidean coordinates in E^3. Then x,y, and z are differentiable and

$$\text{rank}\begin{pmatrix} x_u & y_u & z_u \\ x_v & y_v & z_v \end{pmatrix} = 2 .$$

5) If $r \in N(p) \cap N(q)$ where $N(p)$ is homeomorphic to a disk in the u-v-plane and $N(q)$ is homeomorphic to a disk in the $\bar{u}$-$\bar{v}$-plane, then the natural map $\bar{u} = \bar{u}(u,v)$, $\bar{v} = \bar{v}(u,v)$ determined in $N(p) \cap N(q)$ is differentiable in both directions and therefore:

$$\frac{\partial(\bar{u},\bar{v})}{\partial(u,v)} \neq 0 .$$

1.2 Theorem. (Jordan-Brouwer) If S is a simple, closed surface in E^3, then $E^3 \setminus S = I \cup E$ where I and E satisfy

1) I and E are open, connected, non-empty sets

2) $I \cap E$ is empty

3) I is bounded

4) E is not bounded

I is called the *interior* of S and E is called the *exterior* of S .

1.3 Theorem. If S is a simple, closed surface in E^3 , then S is homeomorphic to a sphere with g handles $(g \geq 0)$ as illustrated below. The number g is called the *genus* of the surface.

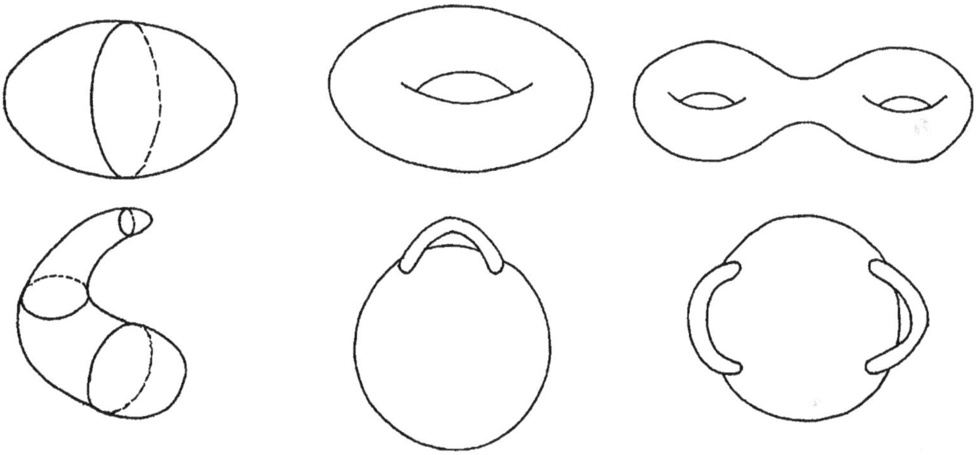

Surfaces with zero handle — Surfaces with one handle — Surfaces with two handles

2. Abstract Closed Surfaces

2.1 Definition. An *abstract closed surface* is a Hausdorff space S satisfying the second axiom of countability and such that conditions 1), 2) and 3) of 1.1 are satisfied. If in addition, condition 5) of 1.1 is satisfied, then the surface is called *differentiable*.

Let S be a differentiable surface and let f be a real-valued function on S . Let $p \in S$ and let N(p) be homeomorphic to a disk in the u-v-plane with p corresponding to (0,0) . Then f is *differentiable* at p if f(u,v) is differentiable at (0,0) . This notion is clearly invariant under a differentiable change of coordinates.

2.2 Definition. An abstract closed, differentiable surface is called *orientable* if all of the parameter disks can be chosen so that all of the Jacobians of condition 5) of 1.1 are positive.

If the surface is a subset of a Euclidean space then an equivalent condition is that it be possible to assign a coherent orientation to the tangent planes to the surface. Since an orientation in a tangent plane at p determines an orientation in a neighborhood of p, this is equivalent to requiring that the orientations assigned in the intersection of two neighborhoods are the same.

2.3 Theorem. A simple closed surface in E^3 is orientable.

Proof [Sketch]: It is easy to see heuristically that this must be true, since, by 1.2, at each point of the surface we can distinguish the normal directed towards the interior. If a definite orientation is chosen for E^3, then at each point the tangent plane can be oriented so that this orientation and the direction of the interior normal gives the chosen orientation of E^3. This process clearly gives a coherent orientation of the whole surface.

2.4 Theorem. An orientable abstract closed surface is homeomorphic to a simple closed surface in E^3, and hence, by 1.3, is homeomorphic to a sphere with g handles $(g \geq 0)$. However, as the following two examples illustrate, there are non-orientable abstract closed surfaces.

2.5 Example. The real projective plane. Let (u,v,w) be an orthogonal coordinate system in E^3 and let p be the point $(0,0,0)$. Let L be the plane $w = 1$. Then each straight line through p, not parallel to L determines exactly one point on L, and conversely. In fact, (u,v,w) may be considered as homogeneous coordinates of the point

$$\left(\frac{u}{w}, \frac{v}{w}, 1\right) \in L .$$

Clearly, the lines parallel to L correspond to the "points at infinite" in the projective L plane; i.e., the points with homogeneous coordinates $(u,v,0)$. Hence the real projective plane is in 1-1 correspondence with the bundle of straight lines through a point in E^3. This bundle of straight lines has a natural 1-1 correspondence with the pairs of antipodal points on the 2-sphere. The set of these pairs of points is in 1-1 correspondence with the upper hemisphere where the antipodal points on the bounding, equatorial circle are identified.

In the above construction, if the unit sphere is taken as the 2-sphere with center at p, then a point on the projective L plane with homogeneous coordinates (u,v,w) maps into the pair $\left(\frac{u}{n}, \frac{v}{n}, \frac{w}{n}\right)$,

$(-\frac{u}{n}, -\frac{v}{n}, -\frac{w}{n})$, where $n = \sqrt{u^2 + v^2 + w^2}$, under the first correspondence; and into $(\frac{u}{n}, \frac{v}{n}, \frac{|w|}{n})$ if $w \neq 0$ or $(\frac{u}{m}, \frac{v}{m}, 0)$ and $(-\frac{u}{m}, -\frac{v}{m}, 0)$, where $m = \sqrt{u^2 + v^2}$ if $w = 0$, under the second correspondence.

This last set is clearly homeomorphic to a closed disk in the plane with diametrically opposite boundary points identified. For neighborhoods of points interior to the disk, we take ordinary Euclidean neighborhoods. For boundary points, two half disks are taken as illustrated.

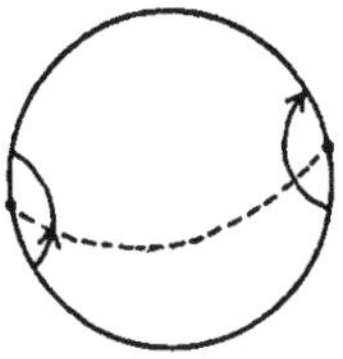

Using this last model, it is clear that the projective plane is non-orientable, since if an orientation is chosen at the center, then the orientation is determined at all interior points. However, because of the identifications of boundary points, each point on the boundary of the disk must have opposite orientations, depending on the direction from which the boundary is approached.

2.6 Example. The Klein Bottle. A Klein bottle can be represented as a rectangle with sides identified as in figure 1) . Figure 2) is a picture of a model of a Klein bottle in E^3 with self-intersections.

1)

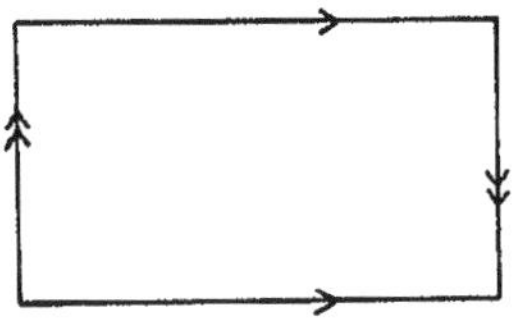

2)

2.7 Theorem. A non-orientable abstract closed surface is homeomorphic to a real projective plane with p handles $(p \geq 0)$ or a Klein bottle with p handles $(p \geq 0)$.

3. General Closed Surfaces in E^3

3.1 Definition. Let S_o be an abstract closed surface with a differentiable structure. Let X be a differentiable map of S_o into E^3 ; i.e.,

$$X(p) = (x(p), y(p), z(p))$$

where $x(p), y(p)$, and $z(p)$ are differentiable functions on S_o, such that, with respect to local parameters u and v,

$$\text{rank}\begin{pmatrix} x_u & y_u & z_u \\ x_v & y_v & z_v \end{pmatrix} = 2 .$$

Then X is called a <u>general closed surface</u> in E^3. Strictly speaking, a general closed surface is an ordered couple $\{S_o, X\}$. However, if $S = X(S_o)$ is the image of S_o under X, we will also call S the general closed surface when no confusion will arise from this convention.

The condition on the rank implies that X is locally 1-1 ; however, the image may have self-intersections. If S_o is non-orientable then the image will necessarily have self-intersections, by 2.3.

<u>3.2</u> <u>Theorem</u>. Every non-orientable closed surface can be imbedded differentiably and locally 1-1 as a general closed surface in E^3.

<u>Proof</u>: Illustration 2) of 2.6 gives such an imbedding of the Klein bottle in E^3. The Boy surface is an imbedding of the projective plane in E^3. For a description of this surface, see Hilbert-Cohn-Vossen: Geometry and the Imagination. By 2.7, all others are these with handles.

<u>3.3</u> <u>Problem</u>. The above imbeddings are geometrical of necessity since no explicit formulas are known for the imbedding of the Klein bottle and the projective plane in E^3. For the projective plane, this could be done, for example, by specifying three functions on the 2-sphere which are even and such that the rank of their Jacobian is 2 . It is conjectured that this can be done with homogeneous forms of even degree. However, it is impossible with quadratic forms and is conjectured to be impossible with quartic forms.

With respect to imbedding surfaces in E^4, an explicit parametric representation of the projective plane is given in Hilbert-Cohn-Vossen. (However, the equational characterization given there is incorrect as can be easily seen by the fact that (in their notation) the plane $z = t = 0$ also satisfies the given equations.) In general, if M^k is a k-dimensional, non-orientable manifold, then it is impossible to represent M^k in E^n by giving $n-k$ equations of the form

$$f_i = 0 \ , \ i = 1,\dots,n-k$$

with maximal functional rank. The proof consists essentially in observing that if such equations could be given, then their gradients would be independent at every point, and hence determine an orientation of M^k ; which is a contradiction.

3.4 Remark. In the remainder of these lectures we shall restrict our considerations to orientable surfaces.

4. Riemannian Geometry

4.1. Let S_o be an abstract closed surface. Then a Riemannian geometry is determined on S_o by giving a covariant, symmetric, second order tensor, g_{ij} , on S_o , such that the quadratic form $g_{ij}du^i du^j$ is positive definite. By 2.4 and 3.2, since every abstract closed surface can be realized as a general closed surface in E^3 , the induced metric given by considering S_o as a subset of E^3 provides an example of a Riemannian metric for an arbitrary surface S_o . The question then arises are such metrics the only possibilities; i.e., given a Riemann metric g_{ij} on an abstract closed surface S_o , does there exist an isometric imbedding of S_o as a general closed surface in E^3 . The following theorem and example show that the answer is no!

4.2 Theorem. Let S be a general closed surface in E^3 . Then there are points p in S such that $K(p) > 0$.

Proof: Since S is compact, S is contained in a sphere with given center of minimum radius R , with the property that S is tangent to the sphere at at least one point. At this point,

$$\text{the curvature of } S \geq \text{the curvature of the sphere} > 0 .$$

Exercise. Show that $\iint_P K\,dA \geq 4\pi$ for every general closed surface S in E^3 , where $P \subset S$ is the set where $K > 0$.

4.3 Example. Let S_o be the surface of a torus of revolution and let α and β be parameter angles as indicated.

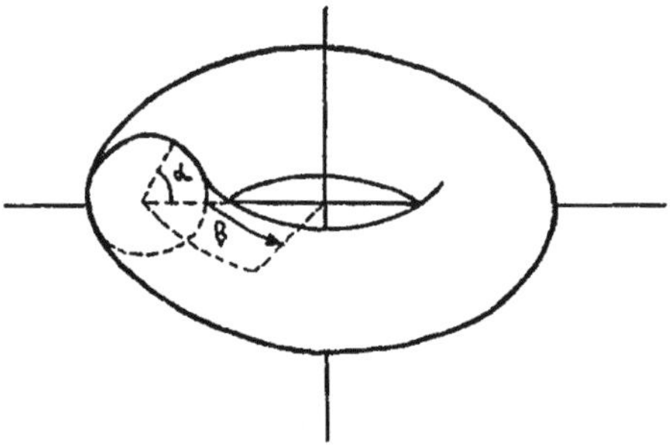

Define
$$ds^2 = d\alpha^2 + d\beta^2 \quad .$$

Then $K \equiv 0$, since in this metric, S_0 is locally like the Euclidean plane. Hence by 4.2, S_0 cannot be imbedded isometrically in E^3. However, S_0 can be isometrically imbedded in E^4; the functions

$$x_1 = \cos\alpha \; , \; x_2 = \sin\alpha \; , \; x_3 = \cos\beta \; , \; x_4 = \sin\beta$$

give such an imbedding since

$$ds^2 = dx_1^2 + dx_2^2 + dx_3^2 + dx_4^2 = d\alpha^2 + d\beta^2 \quad .$$

This counterexample, since it uses the curvature K, is relevant only to the problem of isometries which are three times continuously differentiable; i.e., of class C^3.

4.4. In the above example, we saw that the given metric could be realized in E^4. In general it is known that a k-dimensional, compact, Hausdorff space satisfying the second axiom of countability can be imbedded homeomorphically in E^{2k+1}. If the space in addition is a manifold, then it can be realized in E^{2k}. Hence any abstract closed surfaceface can be realized in E^4. The question remains, can every Riemann metric be realized in E^4? J. Nash has shown that this is true if one is content with a C^1-imbedding. (See Annals of Mathematics, 60 (1954) p. 383-396). His results can be summarized as follows:

If an abstract, closed, differentiable manifold M^n, of dimension n, admits a topological imbedding of class C^∞ in E^k, $k \geq n+2$, then every Riemann metric (of class C^∞) on M^n can be realized on a simple closed manifold of class C^1 in E^k.

Nash has also obtained results on C^∞ isometric imbeddings, (See Annals of Mathematics 63 (1956) p. 20-63), but here the bound on the dimension is much worse. His main theorem here is:

A compact n-manifold with a C^k positive metric has a C^k isometric imbedding in any small volume of Euclidean $(\frac{n}{2})(3n+11)$-space, provided $3 \leq k \leq \infty$.

No definite results have been given for the case $k = 2$.

CHAPTER III

The Total Curvature (Curvatura Integra) of a Closed Surface with Riemannian Metric and Poincaré's Theorem on the Singularities of Fields of Line Elements

1. Singularities of Families of Curves

1.1 Definition. A line element on a surface S is determined by a non-zero tangent vector to the surface. The same line element is determined by all non-zero multiples of the vector. Hence there is no distinguished direction on a line element. Strictly speaking, a line element is a one dimensional linear subspace of the tangent vector space.

A regular (integrable) field of line elements in a region corresponds to a family of curves in the region such that at each point of the region the line element at that point is tangent to the curve through that point.

If a regular field of line elements is given everywhere in a region except at a single point p and if it is impossible to extend the field (uniquely) to p by continuity, then the field is said to have a singularity at p.

1.2 Definition. The index j of an isolated singularity is defined as follows:

Let p be an isolated singularity of a field of line elements and let C be a simple closed curve such that

1) p is the only singularity in the interior of C.
2) There are no singularities on C.

Then the given field induces a field F of line elements on C. Let C be given as a function of a parameter t, $C = C(t)$, $0 \leq t \leq 1$. Choose one of the two possible directions along the line element at $C(0)$. This determines a direction at $C(t)$ for every t, $0 \leq t \leq 1$, by continuity. We wish to measure the total change in angle of this field of directions in going once around C. In order to do this, we must have something to measure angles with respect to. Assume, for the moment, that C is small enough to be contained in the region of validity of a fixed local coordinate system. Within such a region there is defined a field with no singularities; e.g., the lines v = constant. This determines a direction at each point, which will be denoted by U.

Let $\sphericalangle[U,F]$ be the angle between the U direction and the chosen F direction, and let $\delta_C\sphericalangle[U,F]$ be the total change in this angle in going around C once in the positive direction. Then we define

$$2\pi j = \delta_C\sphericalangle[U,F] \quad .$$

It is easy to see that $j = \frac{n}{2}$ where n is an integer.

1.3 Theorem. j does not depend on U or C . Hence the restriction of C to small curves in the definition is unnecessary.

Proof: 1) Let V be another field without singularities. Then

$$\delta_C\sphericalangle[U,F] = \delta_C\sphericalangle[U,V] + \delta_C\sphericalangle[V,F] \quad .$$

However, by choosing C small enough, we can make $\delta_C\sphericalangle[U,V]$ arbitrarily small since they are both continuous fields without singularities. Hence, since $\delta_C\sphericalangle[U,V]$ is an integer multiple of π, $\delta_C\sphericalangle[U,V] = 0$.

2) Since $\frac{1}{\pi}\delta_C$ depends continuously on C , while 2j is an integer, it is clear that j does not depend on C .

1.4 Theorem. If ds^2 is a Riemannian metric and the angle in 1.2 is measured with respect to this metric, then the index doesn't depend on the metric used.

Proof: Let (g_{ij}) and (h_{ij}) be the matrices of the positive definite forms of two Riemannian metrics. Then

$$f_{ij}(t) = (1-t)g_{ij} + th_{ij} \qquad 0 \leq t \leq 1$$

is also positive definite and hence determines a Riemannian metric for each t . But the angles change continuously with t , and 2j is an integer. Hence, j does not depend on the metric.

1.5 Examples. The following examples show that for each $j = \frac{n}{2}$, there is a field with a singularity with index j .

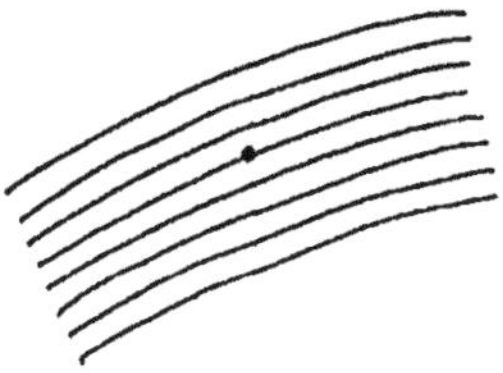

1) no singularity, $j = 0$

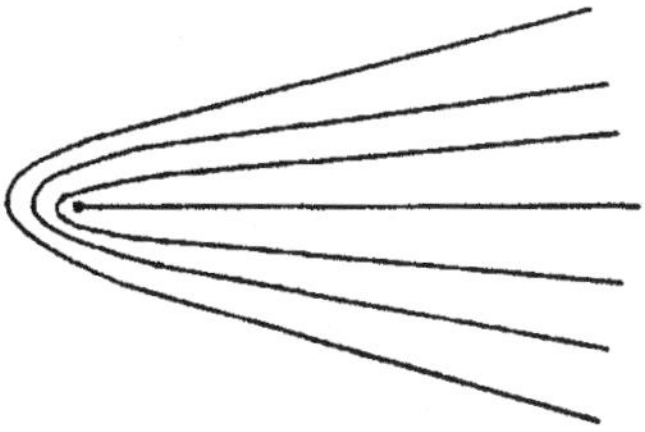

2) $j = \frac{1}{2}$

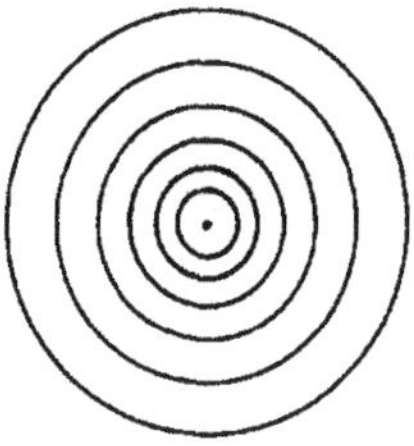

3) $j = 1$

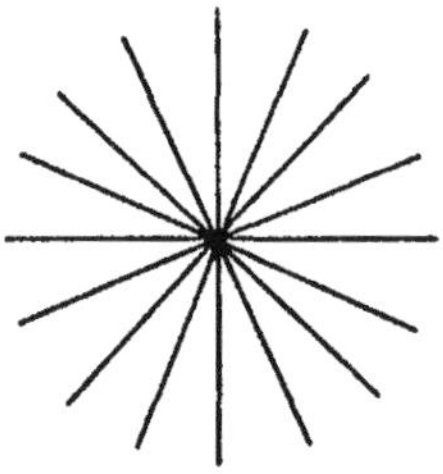

4) $j = 1$

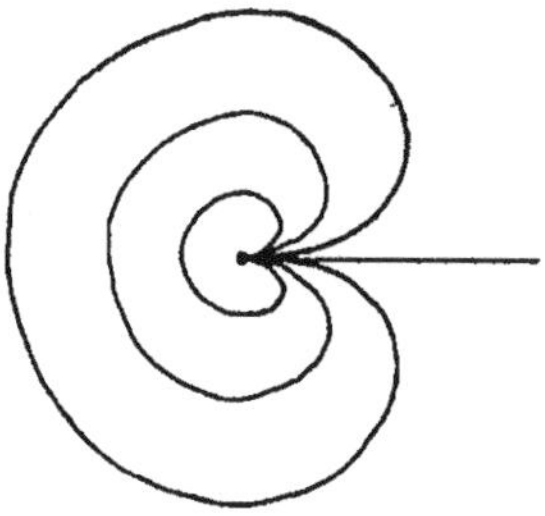

5) $j = 3/2$

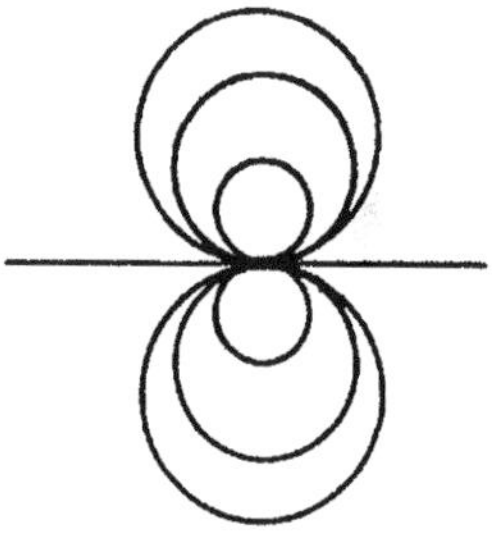

6) $j = 2$

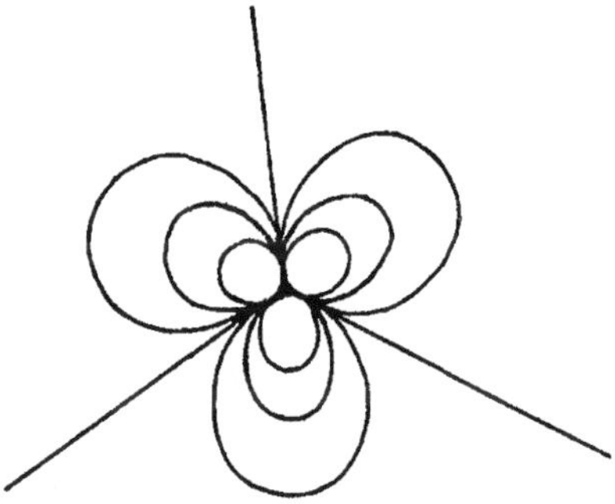

7) $j = 5/2$

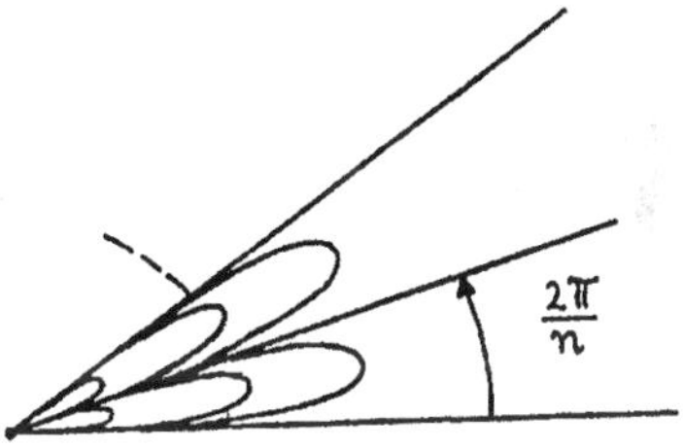

8) $j = \frac{n+2}{2}$, $n \geq 1$

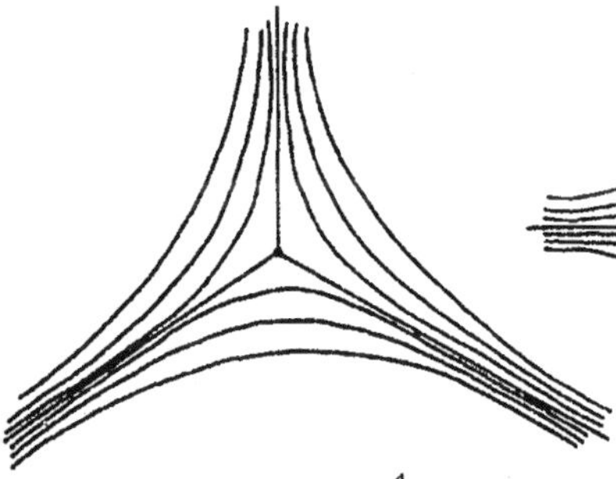

9) $j = -\frac{1}{2}$

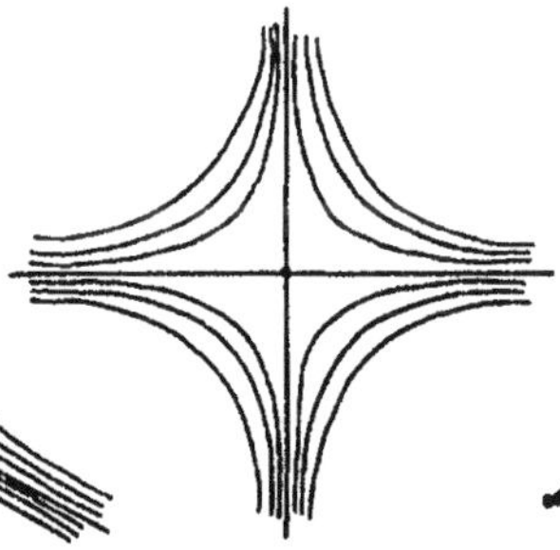

10) $j = -1$

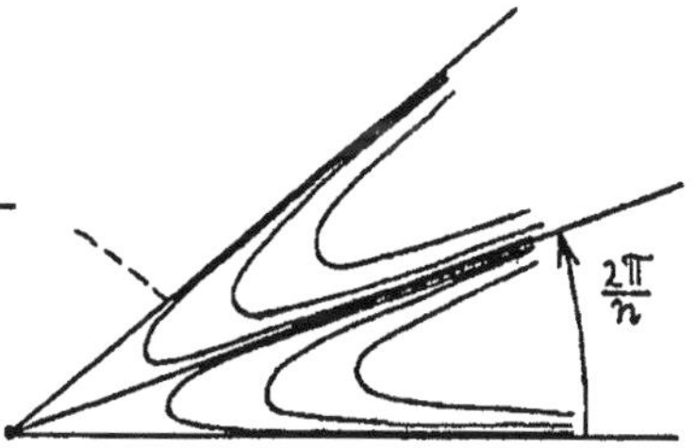

11) $j = \frac{2-n}{2}$, $n \geq 1$

<u>1.6 Theorem</u>. Given a surface of genus g, there is a field of line elements defined on the surface with a finite number of singularities such that the sum of the indices of the singularities is $2-2g$.

<u>Proof</u>: We will sketch the desired differentiable fields with this property.

1a) $g = 0$. Take the great circles through the poles. There are two singularities each of which is like 4) of 1.5 and hence has index $+1$. Therefore $\Sigma j = 2 = 2 - 2\cdot 0$.

1b) $g = 0$. The stereographic projection of a family of parallel streight lines in the plane. There is one singularity at the north pole which looks like 6) of 1.5, and hence has index $+2$.

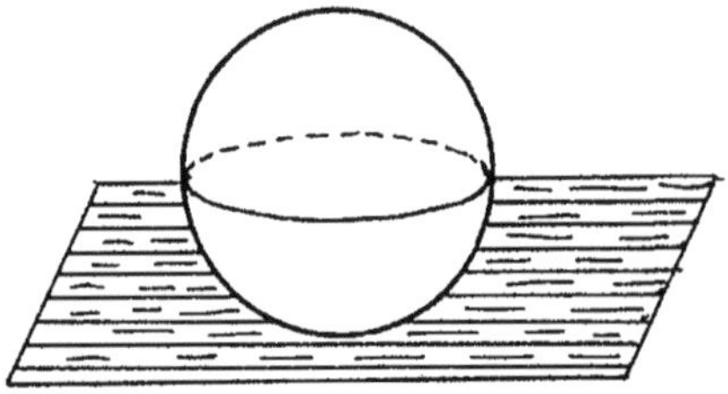

1c) $g = 0$. Take the level lines of z. There are three singularities like 3) of 1.5 with index $+1$, and one like 10) of 1.5 with index -1. Hence $\Sigma j = 2$.

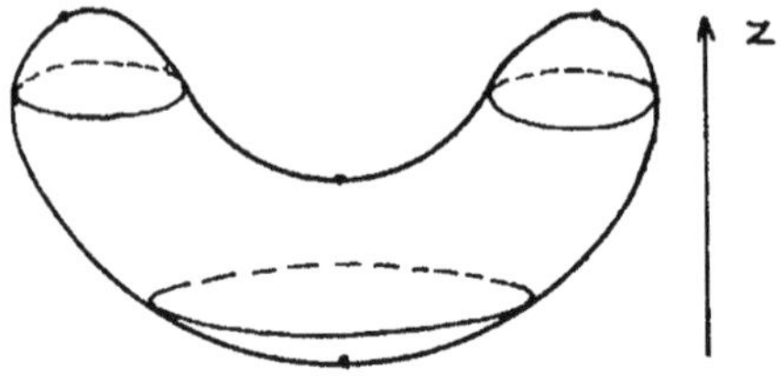

2a) $g = 1$. Take circles of revolution. There are no singularities, and $\Sigma j = 0 = 2 - 2\cdot 1$.

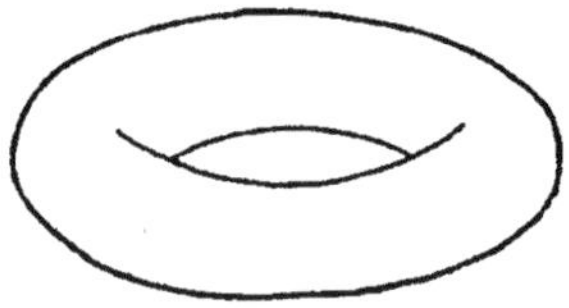

2b) $g = 1$. Take level lines of x. Then there are 2 singularities like 3) of 1.5 with index $+1$ and two like 10) of 1.5 with index -1. Hence $\Sigma j = 0$.

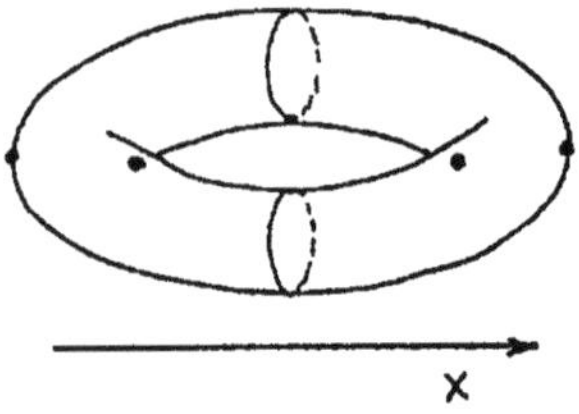

3a) For arbitrary g , if the surface is imbedded as shown, take the level of x . Then there are 2g saddle points like 10) of 1.5 with index −1 and 2 singularities like 3) of 1.5 with index +1 . Hence $\Sigma j = 2 - 2g$.

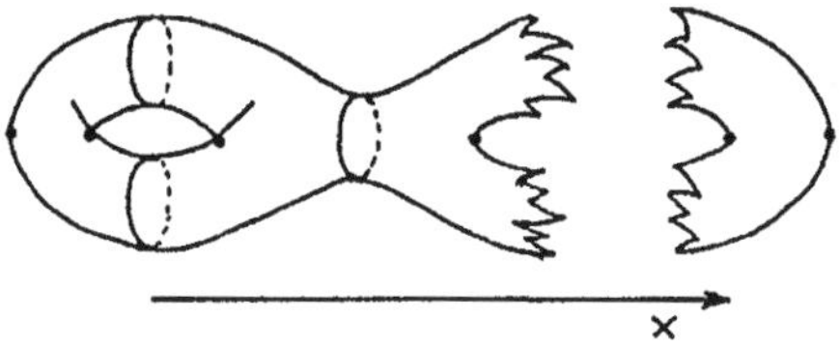

3b) If the surface is represented as two plane figures with boundaries identified as shown, then the indicated field has $2(g-1)$ singularities like 10) of 1.5 with index −1 . Hence $\Sigma j = 2 - 2g$.

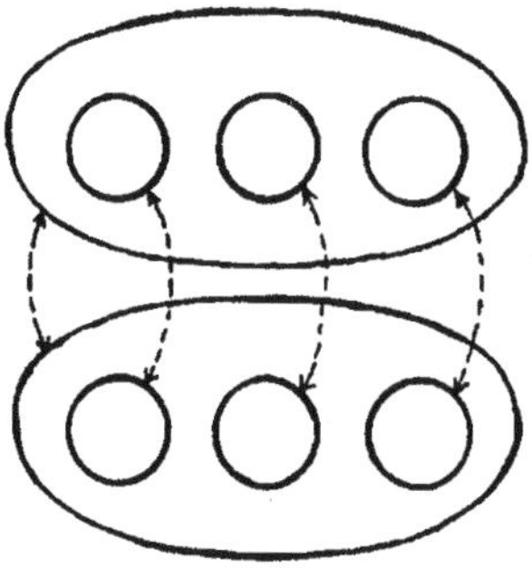

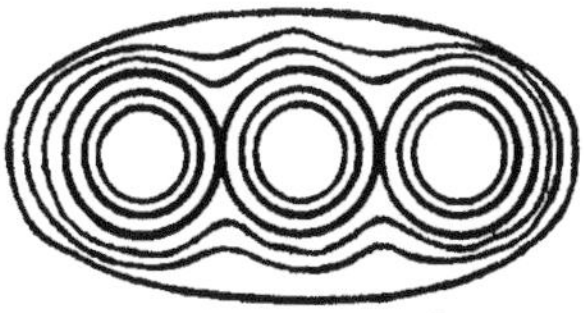

<u>Exercise</u>: Construct a field of line elements with exactly one singularity on a surface of genus $g \geq 2$. Check that the index of the singularity equals $2 - 2g$. (The singularity can be, but is not necessarily, like 11) of 1.5) (p. 209).

<u>1.7 Historical Remark</u>. Poincaré originally considered singularities of differential equations of the form

$$a(u,v)\,du + b(u,v)\,dv = 0 \ .$$

If a and b have a common zero, then the integral curves near the common zero may look like some of the figures of 1.5. For example

a) $u\,du + v\,dv = 0$ gives figure 3)

b) $v\,du + u\,dv = 0$ gives figure 10)

c) $v\,du - u\,dv = 0$ gives figure 4) .

However, it is easy to see that not all the figures of 1.5 correspond to this type of differential equation; for, on an integral curve of $a\,du + b\,dv = 0$, the vector (a,b) is normal to the curve. But (a,b) is a vector with a definite direction. Hence in going around a curve C , it must change by $n \cdot 2\pi$ where n is an integer. Hence the half-integer values for j do not correspond to differential equations.

2. The Main Theorems

2.1 Theorem I: Let S be a closed, orientable surface of genus g with a Riemann metric defined on S so that the curvature K is defined on S. Let there be given a field F of line elements on S with at most a finite number of singularities. Then the sum of the indices j of the singularities of the field is defined and

$$\iint_S K\, dA = 2\pi \Sigma j \quad .$$

Proof: If C is a simple arc on S which does not contain a singularity of F, then F induces a field of directions F_C on C. If Z_C is a parallel field on C, then we define

$$\Phi(C) = \delta_C \sphericalangle [Z_C, F_C] \; .$$

Clearly, given F and given the metric, $\Phi(C)$ does not depend on the chosen parallel field and depends only on C. Since the definition of a parallel field does not depend on the direction in which a curve is traversed, if $-C$ denotes the curve C traversed in the opposite direction, then

$$\Phi(-C) = -\Phi(C) \; .$$

It is a well-known theorem in the theory of surfaces that a surface can be subdivided into 2-cells, where a 2-cell is the topological image of a closed cell in the plane. Clearly, this can be done in such a way that:

1) There are no singularities of F on the boundary of any cell.
2) Each cell contains at most one singularity.
3) Every cell can be covered by a geodesic parameter system.

Call these cells $y_1, y_2 \ldots$, and let $j(y)$ be the index of the singularity in y if there is one and 0 otherwise.

Let y be a fixed cell, let U be the direction of the geodesics in the geodesic coordinate system, and let $b(y)$ be the boundary of y. Then, by 1.2 and 1.3,

1) $$2\pi j(y) = \delta_{b(y)} \sphericalangle [U,F] \; .$$

However, we saw in I, 6.2, that if Z is a parallel field on $b(y)$ then

$$\iint_y K\, dA = \delta_{b(y)} \sphericalangle [U,Z] \quad .$$

This can be rewritten

2) $$-\iint_Y K\, dA = \delta_{b(y)} \sphericalangle [Z,U] .$$

Adding equations 1) and 2) gives

$$2\pi j(y) - \iint_Y K\, dA = \delta_{b(y)} \sphericalangle [Z,F]$$
$$= \sum_{C_i \varepsilon b(y)} \Phi(C_i)$$

where the sum is taken over all arcs C_i in $b(y)$. If this equation is summed over all 2-cells y_k, then

$$2\pi \Sigma j(y) - \iint_S K\, dA = 0$$

since each arc C_i appears in the boundary of exactly two y_k , once as $+C_i$ and once as $-C_i$.

This is the desired result.

2.2 Theorem II: (Poincaré). If F is a field of line elements on S with at most a finite number of singularities, then

$$\Sigma j = 2 - 2g$$

where g is the genus of S .

Proof: Since $2\pi \Sigma j = \iint_S K\, dA$ and $\iint_S K\, dA$ does not depend on the field, Σj is the same for all fields with at most a finite number of singularities. But in 1.6 we gave an example where $\Sigma j = 2 - 2g$. Hence the equality holds for all such fields.

2.3 Theorem III: (The Curvature Integral). If S is a closed orientable surface of genus g with a Riemann metric, then

$$\iint_S K\, dA = 2\pi(2-2g) .$$

Proof: By 1.6, there exists a field of line elements on S with at most a finite number of singularities. Hence the proof is immediate by 2.1 and 2.2.

2.4 Applications

a) $g = 0$. Then $\iint_S K\, dA = 4\pi$. If P is the set of points where $K > 0$, then $\iint_P K\, dA \geq 4\pi$.

(a_1) Any field on such a surface has at least one singularity; and, if it has at most a finite number of singularities, then at least one singularity has a positive index.

b) $g = 1$. Then $\iint\limits_S K\,dA = 0$. This is the only case where it is possible to define a Riemannian metric such that $K = 0$, and the only case in which it is possible to define a field of line elements without singularities.

c) $g \geqslant 2$. Then there is at least one singularity for any field of line elements. If there are at most a finite number of singularities, then at least one singularity has a negative index.

d) g very large. Then K is negative on most of the surface.

<u>2.5 Euler's Formula</u>. Let S be a closed orientable surface with a given subdivision into 2-cells. Let

a_o = number of vertices
a_1 = number of edges
a_2 = number of 2-cells.

Then $a_o - a_1 + a_2 = 2 - 2g$. (This result will not be needed in what follows.)

<u>Proof</u>: We will construct a field where it is obvious that $a_o - a_1 + a_2 = \Sigma j$, which will prove the theorem. We make a barycentric subdivision of the given subdivision by taking as new vertices the original vertices, an interior point of each edge, and an interior point of each cell. New edges are added as indicated. In each triangle of the barycentric subdivision construct a field as shown, where b_o is an original vertex, b_1 the interior point of an edge and b_2 the interior point of a 2-cell. Then at b_o and b_2 there is a singularity like 4) of 1.5 with index +1 and at b_1 there is a singularity like 10) of 1.5 with index -1 . But each b_o corresponds to an original vertex, each b_1 to an original edge and each b_2 to an original 2-cell.

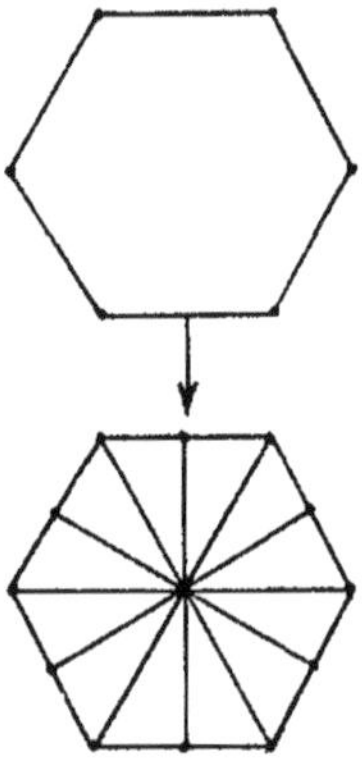

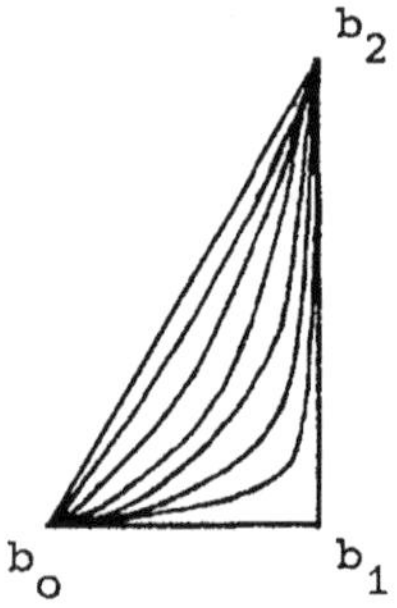

Hence $\Sigma j = a_o - a_1 + a_2 = 2 - 2g$.

Notation: The number $\chi = 2 - 2g$, which appears in the Euler formula, the curvature integral, and the Poincaré theorem, is called the characteristic of the surface.

3. The Degree of the Spherical Map

3.1 Definition: Let $\{S_o, X\}$ be a general, closed, differentiable surface in E^3 (See II, 3.1), where S_o is oriented, and $S = X(S_o)$. Then to each point $p_o \in S_o$, there corresponds a well determined normal of S at the point $p = X(p_o) \in S$. The direction of this normal determines a point $f(p_o)$ on the sphere Σ of directions in E^3 (See I, 7.2). This map $f : S_o \to \Sigma$ is called the spherical map of $\{S_o, X\}$.

Let K be the Gauss curvature and dA the element of surface area of S in the Riemannian metric induced by the imbedding X . Then K and dA may be considered as Gauss curvature and surface element respectively on the parameter surface S_o . Then by I, 7.2, $d\Omega = K\,dA$ is the surface element of $f(S_o)$ on Σ (measured in the ordinary spherical metric of Σ) . The sign of $d\Omega$ is determined by the sign of K since on S_o , we always have $dA > 0$. A region $R_o \subset S_o$ where $K > 0$ is said to cover Σ positively under f . Similarly if $K < 0$, R_o covers Σ negatively. Since the spherical image of the set on S_o where $K = 0$ has area 0 , it follows that

$$\iint_{S_o} K\,dA = \Omega$$

where Ω is the algebraic area of $f(S_o)$ on Σ ; i.e., the sum of the areas covered positively minus the sum of the areas covered negatively. Since $\iint_{S_o} K\,dA = 2\pi(2-2g)$ we get the result

$$\Omega = (1-g)4\pi \quad .$$

Since 4π is the area of Σ , $1-g$ represents the algebraic proportion of Σ that is covered by $f(S_o)$.

3.2 Definition. Let $q \in \Sigma$ satisfy the following conditions:

1) q is the image of only a finite number of points of S_o , $q_1, \ldots, q_m$, under the spherical map f .

2) In a neighborhood of each of these points, f is 1-1 . Then q is said to be in general position with respect to f .

If q is in general position with respect to f , let $P(q)$ be the number of positive coverings of a neighborhood of q and let $N(q)$

be the number of negative coverings. Let $d(q) = P(q) - N(q)$.

3.3 Theorem. If q and q' are in general position then

$$d(q) = d(q') = d .$$

Proof: [Sketch]. To see heuristically that $d(q)$ is constant, we argue as follows:

f essentially defines S_o as a covering surface of Σ and hence $f(S_o)$ may have branch points or fold lines. If q and q' are two points in general position, they may be joined by an arc which avoids the branch points. P and N are constant along this arc except where the arc crosses a fold line. But at such a crossing both P and N either increase by 1 or decrease by 1. Hence $P-N$ remains constant. The theorem below shows that d is independent of the imbedding function X .

3.4 Definition: The number $d = P-N$ is called the degree of f .

Theorem. The degree d of the spherical map of a general closed surface of genus g satisfies $d = 1-g$.

Proof: In 3.1 we saw that $1-g$ is the proportion of Σ covered by $f(S_o)$. But d is clearly also the proportion of Σ covered by $f(S_o)$. Hence

$$d = 1-g .$$

3.5. The definition $d = P-N$ of the degree which we have sketched above can be given quite rigorously in terms of point-set topology. (This definition is the original one given by Brouwer). However, in terms of algebraic topology, a rigorous definition can be given rather easily as follows:

Let M^n and N^n be two n-dimensional manifolds and let f be a continuous map of M^n into N^n . Let $\overline{M^n}$ and $\overline{N^n}$ denote the n-dimensional fundamental cycles on M^n and N^n respectively; i.e., they generate the n-dimensional homology groups of M^n and N^n . Then $f(M^n)$ is a integer is the degree d .

3.6 Theorem. If S_o is a simple closed surface in E^3 , then $f(S_o)$ covers all of Σ . If S_o is a general closed surface and $g \neq 1$, then also $f(S_o)$ covers all of Σ .

Proof: If S_o is a simple closed surface in E^3 , choose the inner normal as the positive normal. Then a given point p on Σ is covered since there is a plane perpendicular to the direction determined by p which just touches the surface from the outside.

If S_0 is a general closed surface, then the above consideration shows that at least one of every pair of antipodal points of Σ is covered. If $g \neq 1$, then $d \neq 0$, and hence every point is covered since if some points were not covered at all, then $P-N = 0$ which is a contradiction.

Remark. The above theorem does not hold if $g = 1$. Consider the surface generated by rotating the curve C about the axis A. This

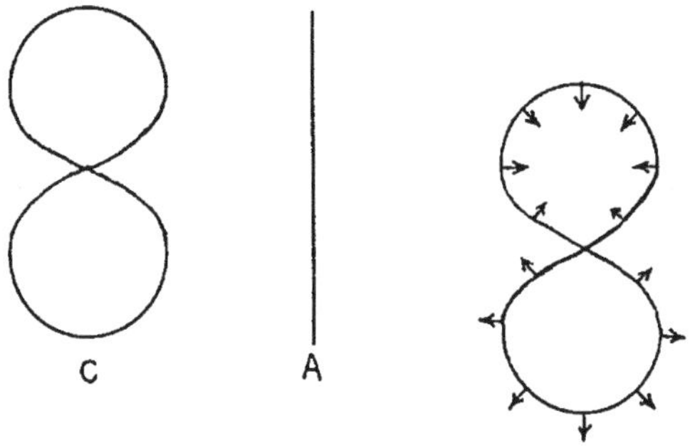

generates a general closed surface in E^3 of genus 1. But, as the figure on the right side illustrates, the normal never points directly upwards, and hence a neighborhood of the north pole is not covered.

4. Generalizations to Higher Dimensions

4.1 The Euler number, the Poincaré number, and the index. Let P^n be an n-dimensional polyhedron, subdivided into simplices. Let a_r be the number of r-simplices and let p_r be the r'th Betti number. Then the Euler number is $\sum_0^n (-1)^r a_r$. The Poincaré number, which generalizes the number $2-2g$, is $\sum_0^n (-1)^r p_r$.

The Euler-Poincaré identity states that

$$\Sigma(-1)^r a_r = \Sigma(-1)^r p_r .$$

This number is called the characteristic χ of P^n.

Furthermore, if P^n is a differentiable manifold then it is possible to consider vector fields on P^n and to define the index of a singularity of such a field. It can be shown that for every field with at most a finite number of singularities, $\Sigma j = \chi$.

4.2 The degree. Let M^n be a general closed, differentiable, oriented, n-dimensional manifold in E^{n+1}. Then the spherical map and the degree d are defined. Then is it true that $d = \frac{1}{2}\chi$? This is false for a circle in the plane, since there, $d = 1$ and $\chi = 0$. Also d is not independent of the imbedding, since for the figure on the right side, $d = 2$.

Theorem. If n is even, then $d = \frac{1}{2}\chi$. If n is odd and > 1, then there exist simple closed manifolds which are homeomorphic but which have different degrees.

4.3 The Gauss Curvature. If n is even and a Riemannian metric is defined on M^n, then a scalar function K can be defined in terms of the first fundamental tensor such that $C_n \int \ldots \int K \, dV = \chi$, where C_n is a constant depending on n. This has been established by Allendoerfer, Weil, and Chern. If n is odd, then $\chi = 0$, so no such formula can exist.

CHAPTER IV

Hadamard's Characterization of the Ovaloids

1. Ovaloids in E^3

1.1 Definition. If p and q are points in E^n, then $\overline{pq}$ denotes the line segment between p and q. A set $S \subset E^n$ is *convex* if for every $p \in S$ and $q \in S$, $\overline{pq} \subset S$. A *convex body* is a compact convex set with a non-empty interior. It is easy to show that a convex body is homeomorphic to a solid sphere (but we will not need this fact). In these notes we will assume in addition that the boundary surface of a convex body in E^3 is several times differentiable.

1.2 Theorem. The surface of a convex body in E^3 satisfies $K \geq 0$.

Proof: Suppose $K < 0$ at some point p of the surface. The one of the principle curvatures, say k_2, is negative (where the inner normal is chosen to be the positive normal in order to determine the signs of the curvatures). Hence the intersection of the surface with the plane determined by the eigenvector corresponding to k_2 and the normal to the surface is a curve with negative curvature. Let q and q' be two points on this curve near p. Then clearly $\overline{qq'}$ is not contained in the convex body; which is a contradiction.

1.3 Definition. An ovaloid is a closed surface which is the boundary of a convex body and which satisfies $K > 0$.

1.4 Theorem (Hadamard). Let S_o be a general closed surface of genus g in E^3 satisfying $K > 0$. Then

1) $g = 0$
2) The spherical map is 1-1 and onto
3) S_o is simple
4) S_o is the boundary of a convex body

Proof: 1) If $K > 0$, then

$$\iint K \, dA = 4\pi(1-g) > 0 \quad .$$

Hence $g < 1$. But since g is a non-negative integer, $g = 0$.

2) We shall give three essentially different proofs that the spherically map is 1-1 and onto.

a) Let $f : S_o \to \Sigma$ be the spherical map and let d be the degree of f. Then

$$d = 1 - g = P - N \; .$$

Since $K > 0$, every point of Σ is in general position. Let $q \in \Sigma$ be covered by f. Then the number of positive coverings is at least one and there are no negative coverings. Therefore $P - N \geq 1$. On the other hand $1 - g \leq 1$. Hence $d = 1$ and every point is covered exactly once. (This gives another proof that $g = 0$.)

b) "The Official Proof". Let $f : S_o \to \Sigma$ be the spherical map. Since $K > 0$, f is a local homeomorphism and hence $f(S_o)$ is an open subset of Σ. On the other hand, since S_o is compact and f is continuous, $f(S_o)$ is a closed subset of Σ. Therefore, since Σ is connected, $f(S_o) = \Sigma$.

To prove f is 1-1, suppose $f(q_o) = f(q_1) = q \in \Sigma$, where $q_o \neq q_1$. Then there is a neighborhood U of q_o such that $f(S_o - U) = \Sigma$. Since there are only positive coverings of Σ, the area of $f(S_o - U)$ is $\geq 4\pi$.

Hence $\iint_{S_o - U} K\, dA \geq 4\pi$, and, therefore,

$$\iint_{S_o} K\, dA > 4\pi$$

which is a contradiction, since by 1),

$$\iint_{S_o} K\, dA = 4\pi \quad .$$

c) This proof is based on the following analogue of the Monodromy Theorem:

Let f be a map of S_o into Σ which is single valued and locally 1-1. Then f is 1-1 in the large and onto. The proof is as follows:

Let $a \in S_o$ and $f(a) = \alpha \in \Sigma$. Since f is locally 1-1, there is a neighborhood $U(\alpha)$ which is in 1-1 correspondence with a neighborhood $U(a)$. Let $\varphi : U(\alpha) \to U(a)$ be this mapping. Then $f \circ \varphi$ is the identity map on $U(\alpha)$. Call φ a "function element" at α. We wish to extend φ to all of Σ.

Let Γ be a curve on Σ from α to β. Then there is a curve C on S_o such that f maps C onto Γ. For suppose there is no such curve. Since φ is 1-1 there is a curve in $U(a)$ which covers $\Gamma \cap U(\alpha)$. Hence there is a first point $\alpha^* \neq \alpha$ on Γ beyond which C does not exist. Let $\{\alpha_i\}$ be a sequence of points on Γ between α and α^* converging to α^*. The α_i correspond to a sequence on C which, by compactness, converges to a point a^*, satisfying $f(a^*) = \alpha^*$.

But by the above argument C can be continued in a whole neighborhood of a^*, which is a contradiction. Therefore, the function φ can be continued along Γ and satisfies $f\circ\varphi$ = identity in a neighborhood of Γ.

If Γ' is another curve from α to β sufficiently close to Γ, then φ can be continued along Γ', given the same function element at β. Hence if Γ'' is any curve from α to β homotopic to Γ, the continuation of φ along Γ'' results in the same function element at β. However, since the sphere is simply connected, all curves from α to β are homotopic. Therefore, the function element at β is independent of the curve Γ. Thus we have defined φ at every point of Σ to satisfy $f\circ\varphi$ = identity. Consequently, f is 1-1 and onto. Note that this proof of statement 2) of our theorem does not make use of the formula $\iint K\,dA = 4\pi$, in contrast to the two other proofs. This is important for Section 2. (For a general treatment of this type of argument, see Chevalley: Theory of Lie Groups, and Steenrod: The Topology of Fibre Bundles.)

3) To prove that the image S of S_o in E^3 is simple, we will show that if T_{a_o} is the tangent plane to S at an arbitrary point $a \in S$ corresponding to $a_o \in S_o$, then there is no point $b_o \neq a_o$ such that the corresponding $b \in S$ is on T_{a_o}. Hence in particular there will be no $b_o \neq a_o$ such that $b = a$ and therefore S is simple.

Let T_{a_o} be the tangent plane at $a \in S$ corresponding to $a_o \in S_o$. Then there is a point $a' \in S$ at a maximum distance from T_{a_o}. Let a_o' be a corresponding point in S_o. Then the tangent plane $T_{a_o'}$ is parallel to T_{a_o} and the normals at a and a' have opposite directions since the spherical map is 1-1. By the same argument, there is no other point $b_o \neq a_o, a_o'$ such that T_{b_o} is also parallel to T_{a_o}.

Now suppose there is a $b_o \neq a_o$ such that b is on T_{a_o}. Then, since T_{b_o} is not parallel to T_{a_o}, it intersects T_{a_o}. But this implies that there are points of S on the opposite side of T_{a_o} from $T_{a_o'}$. Hence by compactness there is a point c on this side of T_{a_o} with maximum distance from T_{a_o}. This is a contradiction since the tangent plane at c must be parallel to T_{a_o}.

4) Since S is simple it has a well defined interior and exterior. Let $a \in S$ and let T_a be the tangent plane to S at a. We have seen in 3) that no point b of S different from a is on T_a. Therefore, S lies entirely on one side of T_a. It also follows from 3) that all

of T_a except a lies in the exterior of S. Suppose for simplicity that T_a is horizontal and S is below T_a. Then from 3) we have in fact that the entire half space above T_a is in the exterior. Now suppose p and q are two points in the interior of S. Then $\overline{pq}$ is below T_a and hence $a \notin \overline{pq}$. Since a is an arbitrary point of S, it follows that no line segment joining two points of the interior of S intersects S. Therefore, the interior of S is convex, so S is the surface of a convex body.

1.5 Remark: In the above theorem the hypothesis can be formally weakened to require only that $K \neq 0$, since we already know that there are points where $K > 0$. In fact if the part 2) of our theorem is suitably modified then it is sufficient to require only that $K \geq 0$ to be able to conclude at least that S is the surface of a convex body.

2. Generalizations to Higher Dimensions

2.1 Theorem. Let M^n be a general closed manifold in E^{n+1}, $n \geq 2$, such that $K > 0$, where K is defined as in two dimensions by the spherical mapping. Then the spherical map is 1-1 and onto and M^n is a simple closed manifold which is the boundary of a convex body in E^{n+1}.

Proof: The proof is the same as in 2) and 3) of 1.4 above, the essential fact being that the n-sphere is simply connected for all $n \geq 2$.

2.2 Remark. The theorem is obviously not true for a curve in the plane (see example). Our proof c) of statement 2) above fails since on the circle S^1, two points α and β may be joined by two curves Γ and Γ' which are not homotopic.

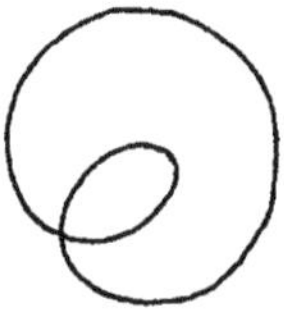

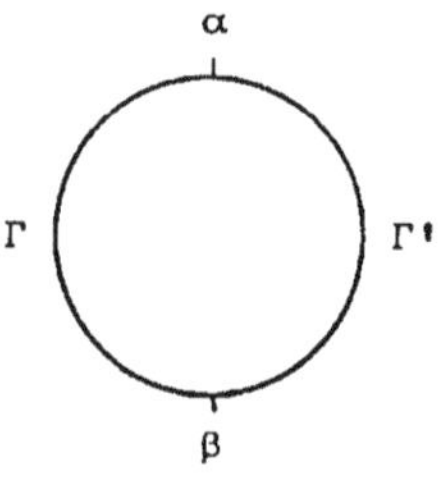

CHAPTER V

Closed Surfaces with Constant Gauss Curvature (Hilbert's Method) - Generalizations and Problems - General Remarks on Weingarten Surfaces

1. A Characterization of the Sphere

1.1 Introduction. Our aim in this section is to prove that the spheres are 1) the only closed surfaces with constant Gauss curvature K, and 2) the only ovaloids with constant mean curvature H. We will actually prove the stronger result that if the principle curvatures k_1 and k_2 of an ovaloid satisfy a relationship $k_2 = f(k_1)$ where f is a decreasing function, then the ovaloid is a sphere. Since $K = k_1k_2$ and $H = \frac{1}{2}(k_1 + k_2)$, the two results, 1) and 2) stated above will follow from this theorem. The difference in the formulation of 1) and 2) is due to the fact that on any closed surface there are points where $K > 0$ (See II, 4.2). Therefore if K is constant, then K is a positive constant and hence by IV, 1.4, the surface already is an ovaloid. The problem of characterizing arbitrary closed surfaces for which H is constant is much more difficult. It will be considered in Chapter VI and VII.

The proof of the above theorem depends on several preliminary lemmas and theorems, the first of which is an important characterization of the spheres.

1.2 Lemma. The spheres are the only closed surfaces for which all points are umbilics.

Proof: Let $\bar{X}$ be the normal to the surface. Then, by I, 8.2,

$$\bar{X}_i = -\ell_i^j X_j \quad .$$

At an umbilic point, $\ell_i^j = k\delta_i^j$, so

$$\bar{X}_i = -kX_i \quad .$$

In terms of coordinates u and v,

1) $\bar{X}_u + kX_u = 0$

2) $\bar{X}_v + kX_v = 0$.

Differentiating 1) with respect to v, 2) with respect to u, and subtracting gives

$$k_v X_u - k_u X_v = 0 \quad .$$

But, since X_u and X_v are independent,

$$k_u = k_v = 0 .$$

Therefore, k = constant.

Hence from 1) and 2) we conclude that

$$\bar{X} + kX = C$$

where C is a constant vector. If $k = 0$, then $\bar{X} = C$ and X is a plane, which is not a closed surface. Thus $k \neq 0$ and

$$\frac{1}{k}\bar{X} = \frac{1}{k}C - X .$$

Finally, since $\bar{X}$ is a unit vector,

$$|X-C'| = \frac{1}{|k|}$$

which is the equation of a sphere of radius $\frac{1}{|k|}$.

Remark. As the proof shows, this result holds in the small. Hence pieces of spheres and pieces of planes are the only possible regions for which all points are umbilics.

1.3 Lemma. Let R be a region of a surface where $K > 0$. Suppose $p \in R$ is not an umbilic point and at p, $k_1 > k_2$. Then it is not possible that k_1 has a maximum at p, and k_2 has a minimum at p.

Proof: Let u and v be parameters such that $F = 0$. Then, by I, 6.6,

$$K = -\frac{1}{2\sqrt{EG}}\left[\left(\frac{E_v}{\sqrt{EG}}\right)_v + \left(\frac{G_u}{\sqrt{EG}}\right)_u\right].$$

This can be rewritten as

1) $$-2(EG)K = E_{vv} + G_{uu} + mE_v + nG_u$$

where m and n are some bounded functions. Since p is not an umbilic point, the parameters u and v can be chosen so that the lines v = constant correspond to the lines of curvature given by k_1 and u = constant to those of k_2. Then $M = 0$ also. In this coordinate system the Codazzi equations are

2) $$L_v = \frac{E_v}{2}\left(\frac{L}{E} + \frac{N}{G}\right) = \frac{E_v}{2}(k_1 + k_2)$$

3) $$N_u = \frac{G_u}{2}\left(\frac{L}{E} + \frac{N}{G}\right) = \frac{G_u}{2}(k_1 + k_2) .$$

Now, in general, if (du, dv) is a tangent direction, then the curvature k in that direction is given by

$$k = \frac{L\,du^2 + 2M\,du dv + N\,dv^2}{E\,du^2 + 2F\,du dv + G\,dv^2} .$$

Since the directions $du = 0$ and $dv = 0$ correspond respectively to k_2 and k_1, we have

$$k_1 = \frac{L}{E} \quad \text{and} \quad k_2 = \frac{N}{G} .$$

The first relation says $L = Ek_1$. Hence differentiation gives

$$L_v = E_v k_1 + E(k_1)_v .$$

By 2) above,

$$E_v k_1 + E(k_1)_v = \frac{E_v k_1}{2} + \frac{E_v k_2}{2} .$$

Thus

$$E(k_1)_v = \frac{E_v}{2}(-k_1 + k_2) .$$

Or

$$E_v = -\frac{2}{k_1 - k_2} E(k_1)_v .$$

Similarly

$$G_u = \frac{2}{k_1 - k_2} G(k_2)_u .$$

Substituting these relations in equation 1) gives

$$-2EGK = -\frac{2E}{k_1 - k_2}(k_1)_{vv} + \frac{2G}{k_1 - k_2}(k_2)_{uu}$$

$$+ m'(k_1)_v + n'(k_2)_u .$$

Or

$$-(k_1 - k_2)EGK = -E(k_1)_{vv} + G(k_2)_{uu}$$

$$+ m''(k_1)_v + n''(k_2)_u .$$

Since $K > 0$ and $(k_1 - k_2) > 0$, the left side of the above equation is negative (and not zero). On the other hand if we have a maximum of k_1, then

$$(k_1)_v = 0 \quad \text{and} \quad -E(k_1)_{vv} \geq 0$$

and if we have a minimum of k_2 then

$$(k_2)_u = 0 \quad \text{and} \quad G(k_2)_{uu} \geq 0 .$$

Hence if both occur simultaneously, the right hand side of the above equation is non-negative; which is a contradiction.

<u>1.4</u> <u>Theorem</u>. Let S be an ovaloid such that there is a point $p \in S$ satisfying

1) $k_1 \geq k_2$
2) k_1 has a maximum at p
3) k_2 has a minimum at p .

Then S is a sphere.

Proof: Since S is an ovaloid, $K > 0$, and hence by Lemma 1.3, p is an umbilic point. Therefore,

$$k_1(p) = k_2(p) .$$

But for all $x \in S$, by hypothesis

$$k_1(p) \geq k_1(x) \geq k_2(x) \geq k_2(p) = k_1(p) .$$

Hence $k_1(x) = k_2(x)$. Therefore, all points are umbilics and consequently, by Lemma 1.2, S is a sphere.

1.4' Theorem. The above theorem can be formulated as follows: If S is an ovaloid which is not a sphere and if

1) $k_1 \geq k_2$
2) k_1 has its maximum at p
3) k_2 has its minimum at q

then $p \neq q$.

1.5 Theorem. Let S be an ovaloid such that $k_2 = f(k_1)$ where f is a decreasing function of k_1 . Then S is a sphere.

Proof: If k_1 has a maximum at p then k_2 has a minimum at p since f is decreasing. Hence by 1.4, S is a sphere.

1.6 Historical Remark. The original problem in this connection was to show that the surface of a sphere is rigid; i.e., it is not possible to "bend" a sphere without changing lengths. Since K is invariant under bendings, this fact is an easy consequence of our theorem that the spheres are the only closed surfaces with constant K . Liebmann gave the first proof of this in 1899. A short time after that Hilbert gave another proof in which he showed that on a closed piece of a surface with constant positive K , which is not a piece of a sphere, if $k_1 > k_2$, then the maximum of k_1 and the minimum of k_2 must lie on the boundary. Our Lemma 1.3 is only a slight generalization of Hilbert's principle lemma. (See the appendix to Hilbert: Grundlagen der Geometrie).

Liebmann proved also (1900) that the spheres are the only ovaloids with constant H . Our theorem 1.5 is included in papers by A.D. Alexandrov (1938) and S.S. Chern (1945).

H. Weyl also proved (1916) a lemma similar to our Lemma 1.3. He showed that on surfaces with $K > 0$, it is not possible for H to have a maximum and K a minimum at the same point. This is an easy consequence of Lemma 1.3.

1.7 Exercise. Give an example of a surface, not a sphere, with $K =$ constant > 0, $k_1 > k_2$, and such that in some interior point of a region of the surface, k_1 has a minimum, and therefore, k_2 a maximum.

Hint: Consider surfaces of revolution with constant K (see Struik for examples). On the equator there are such points.

2. Weingarten Surfaces

2.1 The Curvature Diagram. Let S be a region of a surface. Then at each point $p \in S$, the principle curvatures are uniquely defined by the

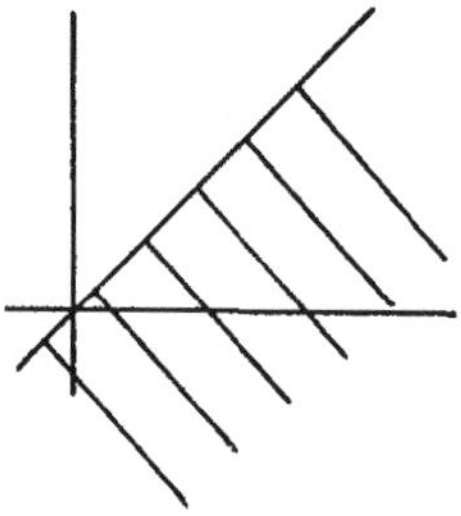

requirement that

$$k_1(p) \geq k_2(p)$$

(i.e., $k_1(p) = H + \sqrt{H^2 - K}$, $k_2(p) = H - \sqrt{H^2 - K}$).
Hence the functions k_1 and k_2 map S into the indicated closed half plane below the main diagonal in the $k_1 - k_2$ - plane. We call the image of S under this mapping the curvature diagram of S. Section 1 above gives some information about the curvature diagrams of surfaces with $K > 0$.

1) A segment of the diagonal line is not a possible curvature diagram since points which map into the diagonal are umbilics. But, by 1.2, the only such surfaces are pieces of spheres for which the curvature diagram is a single point.

2) The cases illustrated in a), b), and c) are not possible for surfaces with $K > 0$ since in each case there is a point where k_1 has a maximum and k_2 has a minimum, contradicting Lemma 1.3. Case c) gives k_2 as a decreasing function of k_1 which is forbidden by Theorem 1.5.

a)

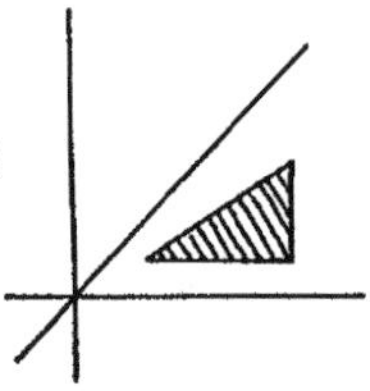

b)

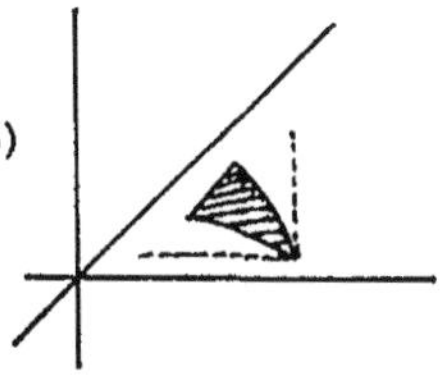

c)

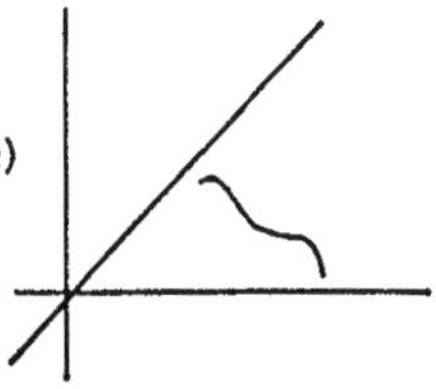

3) The level lines of H are lines perpendicular to the main diagonal, while the level lines of K are hyperbolas with the main diagonal as axis. Hence Weyl's result, quoted in 1.6 and illustrated in d), is included in our results as can be easily seen from figures a) and d).

d)

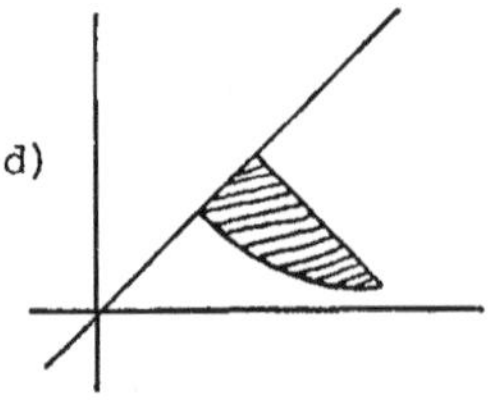

It should be remarked that general sufficient conditions for an arbitrary point-set to be the diagram of a surface are not known.

2.2 Definition. A Weingarten surface (or W-surface) is a surface whose curvature diagram is a curve given by an equation $W(k_1,k_2) = 0$. We will assume that W is differentiable. Since k_1 and k_2 are functions of K and H, $W(k_1,k_2) = 0$ implies that there is a relation $U(K,H) = 0$. However, because differentiability of W with respect to k_1 and k_2 does not imply differentiability of U in the points $k_1 = k_2$, we make the additional assumption that U is also differentiable.

Example. A surface of revolution is a Weingarten surface, since the image of a meridian curve covers the whole curvature diagram, which is therefore a curve.

Exercise. Show that the curvature diagram of an ellipsoid of revolution is an arc of a curve $k_2 = ck_1^3$, where k_2 is the curvature of a meridian curve and $k_1 = \frac{1}{\varrho}$, where ϱ is the distance from the meridian curve to the axis of rotation along the normal to the curve.

2.3 Question. Are the surfaces of revolution the only ovaloids which are Weingarten surfaces?

2.4 The differential equation of a Weingarten surface: Choose a coordinate system for a region of a Weingarten surface so that the surface is given by

$$z = z(x,y) .$$

Then using the equations for K and H given in I, 10.3, we get

$$U(K,H) = \varphi(r,s,t,p,q) = 0 .$$

Hence the Weingarten relation becomes a second order differential equation for z.

The discriminant of this differential equation is computed as follows: Let $P = 1+p^2+q^2$. Then

$$\varphi_r = U_K \frac{t}{P^2} + U_H \frac{1+q^2}{2P^{3/2}}$$

$$\frac{1}{2}\varphi_s = U_K \frac{-s}{P^2} + U_H \frac{-pq}{2P^{3/2}}$$

$$\varphi_t = U_K \frac{r}{P^2} + U_H \frac{1+p^2}{2P^{3/2}} \quad .$$

Hence

$$\begin{aligned}\varphi_r\varphi_t - \frac{1}{4}\varphi_s^2 &= \frac{1}{P^2}\,[U_K^2 K + U_K U_H H + \frac{1}{4} U_H^2] \\ &= \frac{1}{P^2}\,(U_K k_2 + \frac{1}{2} U_H)(U_K k_1 + \frac{1}{2} U_H) \\ &= \frac{1}{P^2}\, W_{k_1} W_{k_2} \quad .\end{aligned}$$

Since $W_{k_1} dk_1 + W_{k_2} dk_2 = 0$ along the diagram curve $W = 0$, the sign of $W_{k_1} W_{k_2}$ is opposite to the sign of the differential quotient $\frac{dk_2}{dk_1}$. Therefore, if $\frac{dk_2}{dk_1} < 0$, then $\varphi_r\varphi_t - \frac{1}{4}\varphi_s^2 > 0$ and the equation is elliptic, while if $\frac{dk_2}{dk_1} > 0$, then $\varphi_r\varphi_t - \frac{1}{4}\varphi_s^2 < 0$ and the equation is hyperbolic. Strictly speaking, the function $z(x,y)$ of our surfaces is an elliptic solution of $\varphi = 0$ if $\frac{dk_2}{dk_1} < 0$ along the diagram curve, and a hyperbolic solution if $\frac{dk_2}{dk_1} > 0$. Under the hypothesis of Theorem 1.5, if we add the condition that f is differentiable and $f' < 0$, we are therefore in the elliptic case. These remarks may show that in any case the sign of $\frac{dk_2}{dk_1}$ on the curve $W = 0$ plays an important role for the properties of a Weingarten surface.

It should be remarked that in terms of parameters u and v , all Weingarten surfaces are characterized by the equation $\frac{\partial(K,H)}{\partial(u,v)} = 0$.

2.5 Closed Analytic Weingarten Surfaces

a) The sphere is a closed analytic Weingarten surface.

b) The surfaces of revolution illustrated below are closed analytic Weingarten surfaces.

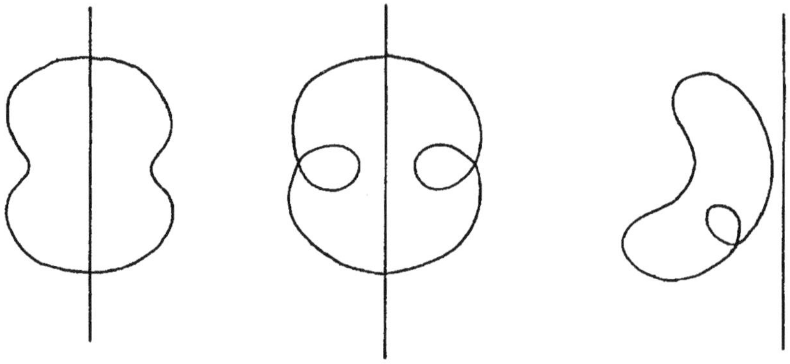

These surfaces are of genus 0 or 1 .

c) Let C be a curve in space. At each point of the curve, put a disk of fixed radius r orthogonal to the curve, with the center of the disk on the curve. The surface generated this way is called a tube. If C is closed, analytic and r sufficiently small, then the tube on C is a closed analytic Weingarten surface. It is of genus 1. The diagram of a tube is illustrated below.

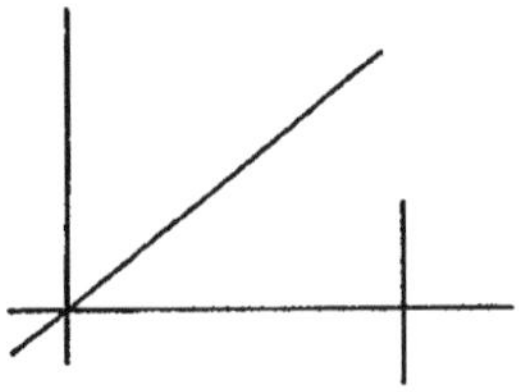

It is not known if these are the only closed analytic Weingarten surfaces.

Exercise. Show that the tubes are the only surfaces with one principle curvature k a constant.

2.6 Examples of C^∞ closed Weingarten surfaces

a) A surface of genus g can be constructed by gluing g handles to a sphere as illustrated

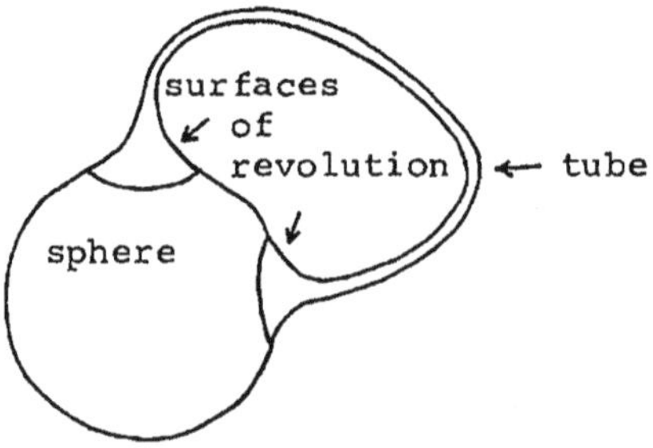

b) A non-orientable surface can be constructed as illustrated

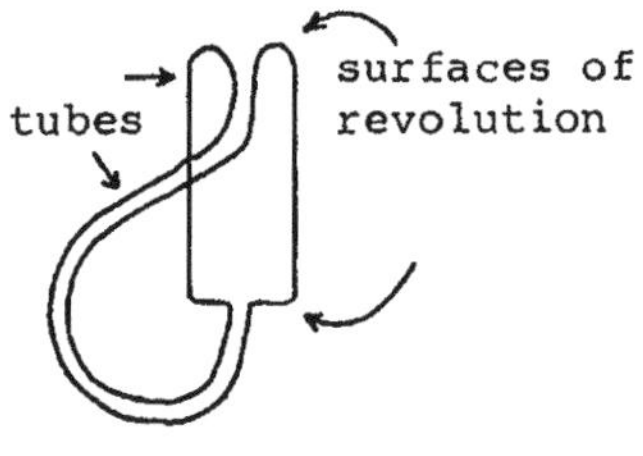

c) One can also have surfaces as illustrated

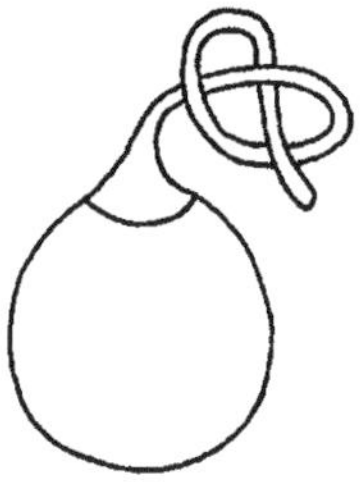

2.7. The most interesting and important case of the general problem of what types of W-surfaces exist is the question whether there exist closed W-surfaces with constant mean curvature H which are not spheres. In the language of the curvature diagram the questions reads as follows: Are there closed W-surfaces whose diagrams are straight line segments

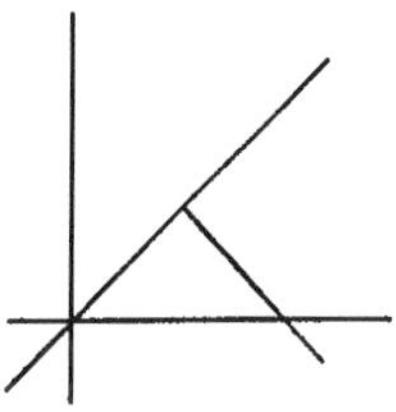

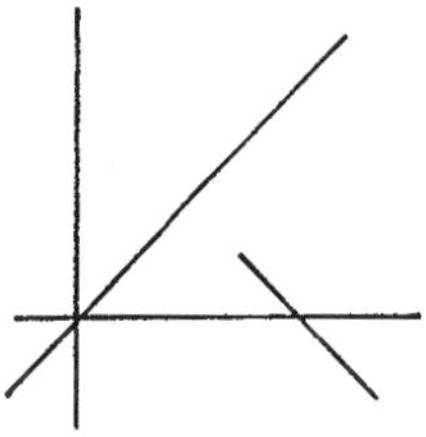

perpendicular to the diagonal $k_1 = k_2$? The answer to this question is certainly "no" if we restrict the investigation to ovaloids; this is a special case of Theorem 1.5 above.

We shall prove in Chapters VI and VII the following two theorems:

1) The only (general) closed surfaces of genus 0 with $H = C$ are the spheres.

2) The only simple closed surfaces (of arbitrary genus) with $H = C$ are the spheres.

The question whether there exist closed surfaces of genus ≥ 1 with $H = C$ and with self-intersections (i.e., not simple) which are not spheres remains unanswered.

Before we enter into Chapters VI and VII, we wish to justify the statement that the knowledge of all closed surfaces with constant H really would be important. We shall in fact show in the following paragraph that this problem is closely related to a classical chapter of geometry; namely, the isoperimetric theorem.

3. The Isoperimetric Problem and Surfaces with Constant H

3.1 Introduction. The isoperimetric problem in two dimensions is to find the shortest simple, closed curve enclosing a fixed area. The solution is a circle.

The analogous problem in three dimensions is to find the region of fixed volume with minimal surface area. Here again the classical answer is the sphere. This problem is related to the discussion of surfaces with constant H as follows: We will prove that a surface has constant mean curvature if and only if its area A is stationary (in a sense to be defined) with respect to volume preserving variations. Hence the conjecture that all simple closed surfaces with constant mean curvature are spheres is equivalent to the conjecture that A, considered as a function on the set of all simple closed surfaces enclosing a fixed volume, has exactly one stationary value; namely, its absolute minimum.

3.2. Let $V(S)$ denote the volume of the interior of a simple closed surface S. Let $\mathcal{S}$ be the collection of all simple closed surfaces S such that $V(S) = 1$. Let $A(S)$ be the area of S. Then A is a function on $\mathcal{S}$ which has exactly one absolute minimum; namely, when S is a sphere. Let S be a fixed surface and consider a one parameter family of continuous and differentiable variations of S, indexed by a parameter t. Let S_t denote the varied surface. Then we require that $S_o = S$ and that for each t, $S_t \in \mathcal{S}$. These variations are called volume preserving variations. Let $A(t) = A(S_t)$. Then A is a differentiable function of t. If $A'(0) = 0$ for all volume preserving variations, then S is called a stationary surface.

We shall prove in 3.4 that a simple closed surface is stationary if and only if its mean curvature H is constant.

3.3 Some Formulas. Let S be a surface given by the vector X and let $X(t)$ be a variation of S where $X(0) = X$. Let $\varphi = X'(0)\bar{X}$ denote the normal component of the variation vector $X'(0)$. We indicated in I, 8.7 that

1) $A'(0) = -2 \iint \varphi H \, dA$.

This was an immediate consequence of the formula

1') $A'(0) = -2\iint \varphi\, H\, dA + \oint(\bar{X}, X', dX)$

which holds for surfaces with boundaries.

Similarly, it can be shown that

2) $V'(0) = -\iint \varphi\, dA$.

This is a consequence of the general formula

2') $V'(0) = -\iint \varphi\, dA + \frac{1}{3}\oint(X', X, dX)$

where the volume V is given by

3) $3V = -\iint X\bar{X}\, dA$.

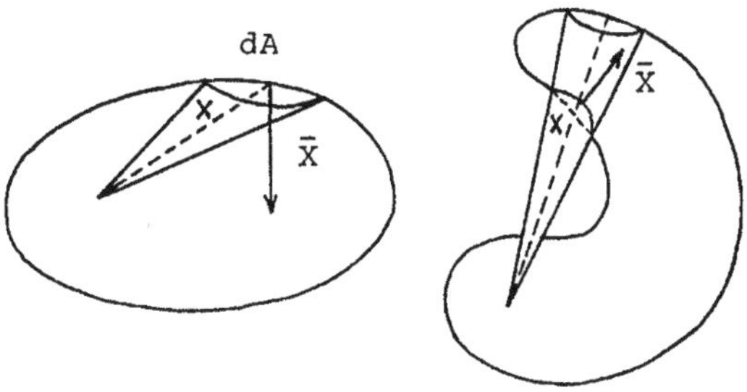

Equation 3) can be derived by considering the figure at the left. $-X\bar{X}$ is the height of the cone of base dA . Hence $-X\bar{X}\, dA$ is 3 times the volume of this cone. Therefore $3V = -\iint X\bar{X}\, dA$. For a non-convex region, the right figure indicates how the proof is carried out.

Exercise. Derive formulas 2') and 2) above using formula 3).

3.4 Theorem. Let S be a simple closed surface. Then S has constant mean curvature H if and only if S is a stationary surface.

Proof: Let S be given by the vector X and suppose for simplicity that $V(S) = 1$. Sufficiency is trivial; for, suppose H is constant and $X(t)$ is a volume preserving variation of S . Then

$$V'(0) = -\iint \varphi\, dA = 0$$

and hence

$$A'(0) = -2\iint \varphi\, H\, dA = -2H\iint \varphi\, dA = 0 .$$

Conversely, suppose $A'(0) = 0$ for every volume preserving transformation. Then we must show that H is constant. Let φ be an arbitrary function defined on S such that $\iint \varphi\, dA = 0$. We wish to show first that then φ is in fact the normal component of a volume preserving variation. Consider the family of surfaces

$$X_1(t) = X + t\varphi \bar{X} .$$

Let $V_1(t)$ denote the volume of $X_1(t)$. Then

$$V_1(0) = V(S) = 1 .$$

Now the normal component of $X_1'(0)$ is clearly given by $X_1'(0)\bar{X} = \varphi\bar{X}\bar{X} = \varphi$. Hence, by equation 2) of 3.3,

$$V_1'(0) = -\iint X_1'(0)\bar{X}\, dA = -\iint \varphi dA = 0$$

by hypothesis. However, the variation $X_1(t)$ need not be volume preserving. This is remedied by taking the family of surfaces

$$X(t) = V_1^{-1/3}(t) X_1(t) .$$

Then, clearly, by equation 3) of 3.3

$$V(t) \equiv 1 .$$

Hence $X(t)$ is a volume preserving variation of S . Now, since $V_1'(0) = 0$, it follows that

$$X'(0) = X_1'(0) = \varphi\bar{X} .$$

Hence taking the scalar product with the unit vector $\bar{X}$ gives

$$\varphi = X'(0)\bar{X} .$$

Therefore φ is not only the normal component of $X_1'(0)$ but is also the normal component of $X'(0)$; and thus φ is the normal component of a volume preserving variation.

By hypothesis, S is stationary; so

$$A'(0) = -2\iint \varphi H\, dA = 0 .$$

Thus $\iint \varphi H\, dA = 0$.

Also if h is an arbitrary constant

$$\iint \varphi h\, dA = 0$$

and hence for any function φ such that $\iint \varphi dA = 0$ and for any constant h ,

$$\iint \varphi (H-h)\, dA = 0 .$$

Now let h be the mean value of H ,

$$h = \frac{1}{A}\iint H\, dA .$$

Then $\iint (H-h)\, dA = 0$ and consequently (because we may put $\varphi = H-h$)

$$\iint (H-h)^2 dA = 0 .$$

Therefore $H \equiv h$, which concludes the proof.

3.5. The condition that H is constant occurs in another connection; namely, a free soap bubble is in equilibrium (no matter how unstable) if and only if H is constant. Of course, the only experimentally known examples are spheres. But, for example, there may very well be cases of soap bubbles of positive genus which are in equilibrium.

3.6 General Closed Surfaces

In order to discuss Theorem 3.4 for non-simple closed surfaces, it is necessary to generalize the notion of volume. For a closed curve C in the plane, the "order" of a point $x \notin C$ with respect to C is defined to be the algebraic number of times C winds around x. I.e.,

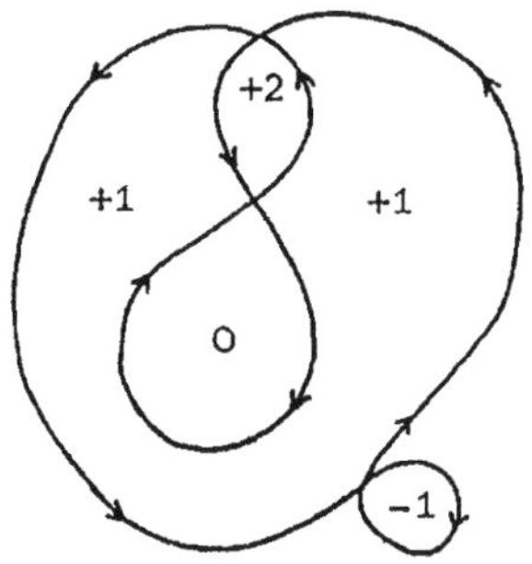

$\mathrm{order}_C(x) = \frac{1}{2\pi}\,\delta_C\theta$ where θ is the angle in a set of polar coordinates with origin at x. Clearly the order depends only on the connected component of the complement of C as indicated in the figure.

For a closed surface $S \subset E^3$ and a point $x \notin S$, take a small sphere about x and project S onto this sphere from x. The degree of this map is defined to be the order of x. As above, the order depends only on the connected component of the complement of S in which the point is located. Hence we can attach an integer d_i to each such connected component R_i. Define

$$V = \Sigma d_i \,\mathrm{Vol}(R_i) \;.$$

Then it can be shown that with this definition of V, again

$$3V = -\iint X\bar{X}\,dA \;.$$

Our Theorem 3.4 and its proof hold also in this situation. T. Rado has shown that the isoperimetric inequality also remains true.

CHAPTER VI

General Closed Surfaces of Genus 0 with Constant Mean Curvature - Generalizations

The main theorem of the chapter has already been formulated in V, 2.7. The proof will be given in Section 2 and some generalizations will be made in Section 3. The first paragraph is devoted to some preparatory formulas and computations.

1. Isothermic Parameters

1.1. In I, 10.2 we discussed the introduction of isothermic parameters; i.e., parameters u, v which satisfy

$$ds^2 = E(du^2 + dv^2) \; .$$

The basic entities in such a parameter system are as follows:

1) $$K = k_1 k_2 = \frac{LN-M^2}{E^2}$$

2) $$H = \frac{1}{2}(k_1 + k_2) = \frac{L+N}{2E} \; .$$

The lines of curvature are given by

3) $$-M\,du^2 + (L-N)\,du\,dv + M\,dv^2 = 0 \; .$$

The Codazzi equations are

$$L_v - M_u = \frac{E_v}{2E}(L+N) = E_v H$$

$$M_v - N_u = -\frac{E_u}{2E}(L+N) = -E_u H \; .$$

But, since $EH = \frac{L+N}{2}$

$$E_v H = -EH_v + \frac{L_v}{2} + \frac{N_v}{2}$$

$$E_u H = -EH_u + \frac{L_u}{2} + \frac{N_u}{2} \; .$$

Hence the Codazzi equations can be written

4) $$\left(\frac{L-N}{2}\right)_u + M_v = EH_u$$

$$\left(\frac{L-N}{2}\right)_v - M_u = -EH_v$$

1.2 Complex Parameters: If u and v are isothermic parameters, we can introduce the complex parameters

$$w = u + iv \; , \qquad \bar{w} = u - iv \; .$$

One verifies easily the rules for differentiation

$$2\frac{\partial}{\partial w} = \frac{\partial}{\partial u} - i\frac{\partial}{\partial v}$$

$$2\frac{\partial}{\partial \bar{w}} = \frac{\partial}{\partial u} + i\frac{\partial}{\partial v}$$

i.e., for an arbitrary complex function $F(w,\bar{w}) = P + iQ$

$$2F_w = (P_u + Q_v) - i(P_v - Q_u)$$

$$2F_{\bar{w}} = (P_u - Q_v) + i(P_v + Q_u) \; .$$

Let $\Phi(w,\bar{w}) = \frac{L-N}{2} - iM$.

Then from 1.1.1 and 1.1.2 it follows that

1) $$\frac{|\Phi|}{E} = \frac{|k_1-k_2|}{2} \; .$$

Hence the umbilic points of a surface S are the zeros of Φ . A simple computation shows that the equation 1.1.3 for the lines of curvature can be written

2) $$\mathrm{Im}\{\Phi(dw)^2\} = 0 \; .$$

This is equivalent to

$$\arg\Phi + 2\arg(dw) = m\pi \quad (m \text{ an integer})$$

or

2') $$\arg dw = \frac{m\pi}{2} - \frac{1}{2}\arg\Phi$$

where dw is the tangent element of a line of curvature.

By multiplying the second equation of 1.1.4 by i and adding it to the first, the Codazzi equations can be written

3) $$\Phi_{\bar{w}} = EH_w \; .$$

1.3 <u>The Index of an isolated umbilic point</u>: Let p be an isolated umbilic point. Then p is an isolated singularity of each of the two families of lines of curvature (the one family corresponding to k_1 and the other to k_2 , where we retain the convention $k_1 \geq k_2$) . Therefore p has an index with respect to each of these families (see III, 1.2); but, because the lines of one family are orthogonal to the lines of the other family, it follows immediately from the definition of the index that these two indices are equal. Therefore the index of an isolated umbilic point is well defined and satisfies

$$j = \frac{1}{2\pi}\delta(\arg dw)$$

where δ means the variation if one goes once around p on a small curve in the positive sense and where dw has the same meaning as in 1.2.2'. From 1.2.2' it follows, since the integer m remains unchanged, that

$$j = -\frac{1}{2\pi}\frac{1}{2}\delta\,(\arg\Phi)\ .$$

1.4 Parameter Transformations: If the isothermic parameters u, v are replaced by other regular parameters x, y, then these new parameters are also isothermic if and only if $z = x+iy$ is an analytic function of $w = u+iv$ with non-vanishing derivative, i.e., we have

$$z = z(w)\ ,\ z' \neq 0\ .$$

This means that the correspondence between the w-plane and the z-plane is conformal.

We are interested in how our function $\Phi(w,\bar{w})$ introduced in 1.2 changes under such a parameter transformation. Now, from the definition of Φ and the definitions of L, M, and N, it follows by an easy calculation that

$$\Phi = -2X_w\bar{X}_w$$

where, as always, X is the position vector and $\bar{X}$ the normal vector of our surface. Similarly, if $\Psi(z,\bar{z})$ denotes the function analogous to $\Phi(w,\bar{w})$ for the parameters x, y, then

$$\Psi = -2X_z\bar{X}_z\ .$$

But, since

$$X_w = X_z\frac{dz}{dw}\ ,\quad \bar{X}_w = \bar{X}_z\frac{dz}{dw}$$

we have

$$\Phi = \Psi\left(\frac{dz}{dw}\right)^2\ .$$

Or

$$\Phi(dw)^2 = \Psi(dz)^2\ .$$

This formula describes the transformation of Φ. In the terminology which is usual in the theory of Riemann surfaces, the rule of transformation may be formulated as follows: With respect to conformal parameter transformations, $w \leftrightarrow z$, $\Phi(dw)^2$ transforms like a complex quadratic differential (where the coefficient Φ is a function of w and $\bar{w}$).

2. The Main Theorem

2.1 Theorem: Let S be a general closed surface of genus 0 (i.e., the parameter surface S_0 is of genus 0) with constant mean curvature H. Then S is a sphere.

We shall give two distinct proofs of this theorem, both of them using the characterization of the spheres given in V, 1.2. Thus we shall prove that all points of the surface S are umbilics. Both proofs will depend on Lemma 2.2 below and the first proof will also require Theorem 2.3.

2.2 Lemma. The condition H = constant is equivalent to the condition that Φ be an analytic function of w (Notation as in 1.2).

Proof: $H = c$ is equivalent to $H_u = H_v = 0$ and hence by 1.1.4 is equivalent to the real and imaginary parts of Φ satisfying the Cauchy-Riemann equations. (The lemma also follows from 1.2.3).

2.3 Theorem. Let R be a region of a surface with constant H and let U be the set of umbilic points. Let $p \in U$. Then

1) either p is an interior point of U
2) or p is an isolated point of U and the index of p is negative.

Proof: By 1.2.1, U is the set of zeros of the function Φ which, by Lemma 2.2, is an analytic function of w. Thus, either $\Phi \equiv 0$ and all points belong to U, or $\Phi \not\equiv 0$ and p is an isolated point of U. In this case we can apply 1.3. Since Φ is analytic

$$\Phi(w) = cw^n + \dots \quad \text{where} \quad c \neq 0 \ , \ n \geq 1$$

and hence $\delta(\arg \Phi) = 2\pi n$.

Consequently,

$$j = -\frac{1}{2\pi} \cdot \frac{1}{2} \delta(\arg \Phi) = -\frac{n}{2} < 0 \ .$$

2.4 First proof of the Main Theorem: We can interpret the lines of curvature and the umbilic points of S as lines and points of S_o. Let U be the set of umbilic points of S_o. Since S_o has genus zero, by Poincaré's theorem (III, 2.2, and III, 2.4 a_1) U is non-empty and if U is finite, then at least one point of U has positive index. Hence by Theorem 2.3, U is infinite. Since S_o is compact, U has a point of accumulation, p. But U is the set of zeros of the continuous function $k_2 - k_1$ and consequently is closed. Hence $p \in U$, and, again by Theorem 2.3, p is an interior point of U. Let U* denote the set of all interior points of U.

Now, suppose there exists a point $q \notin U^*$. Then a continuous path from p to q would have a first common point with the closed non-empty set $S_o - U^*$. This point, being a point of accumulation of U, would belong to U but would be neither an interior point of U nor an isolated point of U, contradicting Theorem 2.3. Therefore q does not exist; i.e. $U^* = S_o = U$, and S is a sphere.

2.5 Second Proof of the Main Theorem: The metric of S induces on S_o a Riemannian metric, and therefore a measure of angle. Thus S induces the structure of an (abstract) Riemann surface on S_o (as defined in the theory of complex analytic functions). On this Riemann surface we have the quadratic differential $\Phi dw^2 = \Psi dz^2$ as discussed in 1.4. By Lemma 2.2, this differential is analytic. Since the zeros of this differential are the umbilic points, the proof would be complete if we could show that $\Phi \equiv 0$. Therefore our main theorem may be considered as a corollary of the following theorem about Riemann surfaces.

2.6 Theorem: On a compact Riemann surface S_o of genus 0 , there exists no analytic quadratic differential Φdw^2 except the trivial one, $\Phi \equiv 0$.

Proof: One way to prove this theorem is to follow exactly the lines of our "first proof" given above in 2.3 and 2.4. One considers on S_o the curves defined by 1.2.2 and their singularities. As in 2.3 one shows that the indices of the singularities are negative and as in 2.4, one finally proves, using Poincaré's theorem, that $\Phi \equiv 0$.

However, the theorem can also be proved without using the Poincaré Theorem and using instead the fact (which is a part of the general uniformization theorem) that there exists only one conformal type of compact Riemann surface of genus 0 . For this reason we may assume that our surface S_o is the ordinary sphere of complex numbers which can be covered by two parameter neighborhoods; one of them in terms of w covering all of the sphere except the point $w = \infty$, and the other in terms of $z = w^{-1}$ covering all of the sphere except the point $w = 0$. The coefficients of the differential $\Phi dw^2 = \Psi dz^2$ are connected by the relation

$$\Phi(w) = \Psi(z)\left(\frac{dz}{dw}\right)^2 = \Psi(z)w^{-4} = \Psi(z)z^4 .$$

But Φ is an entire function of w and Ψ is regular for $z = 0$. Hence $\Phi = 0$ for $w = \infty$. Therefore, $\Phi \equiv 0$ by Liouville's Theorem.

2.7 Remark. The appearance of complex analytic functions in the investigation of surfaces with $H = c$ is not very surprising if one recalls that the class of these surfaces includes the minimal surfaces (defined by $H = 0$). The connections between minimal surfaces and complex analytic functions form a classical chapter of mathematics.

On the other hand, our "main theorem" is trivial in the case $H = 0$, for in this case $K \leqslant 0$ everywhere, which is impossible on a closed surface (see II, 4.2).

3. Special Weingarten Surfaces

3.1 Introduction. In this section we will again study surfaces on which a relation

1) $$W(k_1,k_2) = 0 \qquad (\text{as always } k_1 \geqslant k_2)$$

holds (see V, 2.2). We wish to apply the method which furnished the "first proof" of our main theorem in Section 2 to functions W more general than $W = k_1 + k_2 - c$. Since this method is mainly concerned with the umbilic points, it is natural to impose conditions on W with respect only to their behavior at points where $k_1 = k_2$. We shall always assume that $W(k_1,k_2)$ has continuous first derivatives and that

$$(W_{k_1}, W_{k_2}) \neq (0,0) \quad \text{where} \quad k_1 = k_2 .$$

This means that

2) $$\frac{dk_2}{dk_1} = \kappa \quad \text{exists when} \quad k_1 = k_2$$

(κ may be infinite). The decisive hypothesis is

2') $$\kappa = -1 \quad \text{where} \quad k_1 = k_2 .$$

We will prove that an analytic closed W-surface of genus 0 which satisfies 2') is a sphere.

If, instead of 1), the Weingarten relation is given in the form

1') $$U(K,H) = 0$$

where U is differentiable at the points where $k_1 = k_2$ (or $K = H^2$), then 2) is equivalent to

2*) $$U_K H + \frac{1}{2} U_H \neq 0 \quad \text{where} \quad K = H^2 .$$

This follows immediately from

$$U_{k_1} dk_1 + U_{k_2} dk_2 = 0$$

$$U_{k_1} = U_K k_2 + \frac{1}{2} U_H , \quad U_{k_2} = U_K k_1 + \frac{1}{2} U_H$$

and from the facts that if $k_1 = k_2$, then $U_{k_1} = U_{k_2}$ and $H = k_1 = k_2$. However, we shall not use this form of the statement.

We shall actually use conditions 1), 2), and 2') in the following weaker form: Suppose p_o is an umbilic point such that there is a sequence $\{p_n\}$ of non-umbilic points converging to p_o. Let

$$h = H(p_o) = k_1(p_o) = k_2(p_o) .$$

Then condition 2) implies

2") $$\lim_{p_n \to p_o} \frac{k_2(p_n)-h}{k_1(p_n)-h} = \kappa .$$

Therefore

3) $$\lambda = \frac{1+\kappa}{1-\kappa} = \lim_{p_n \to p_o} \frac{H(p_n)-h}{\frac{1}{2}[k_1(p_n)-k_2(p_n)]}$$

for all sequences $\{p_n\}$ of non-umbilic points converging to p_o. Or, in terms of an isothermal coordinate system at p_o such that $E(p_o) = 1$

3') $$\lambda = \lim_{p_n \to p_o} \frac{H(p_n)-h}{|\Phi(p_n)|}$$

where Φ is the function of 1.2. This last condition is all we will actually use. Since this makes no use of a Weingarten relation, we will in fact prove a more general theorem than was promised. (see 3.5 below).

3.2 Theorem. If S is a general closed analytic, surface of genus 0 satisfying condition 3) with $\lambda = 0$ (i.e., $\kappa = -1$), in all umbilic points, then S is a sphere.

3.3 First part of the proof. Assuming Poincaré's theorem, as in 2.3 it is clearly sufficient to show that if p_o is an umbilic point then either

1) p_o is an interior point of the set of umbilic points; or,

2) p_o is an isolated umbilic point and the index j of p_o is negative. We may assume that p_o is not an interior point of the set of umbilic points. Therefore, there is a sequence $\{p_n\}$ of non-umbilic points which converge to p_o and condition 3) is applicable to this sequence.

Since we are on an analytic surface, Φ and H have Taylor's series developments around p_o.

$$\Phi(w,\bar{w}) = \Phi^{(n)}(w,\bar{w}) + \Phi^{(n+1)}(w,\bar{w}) + \dots, \; n > 0$$

$$H(w,\bar{w}) = H^{(0)}(w,\bar{w}) + H^{(1)}(w,\bar{w}) + \dots$$

where $\Phi^{(k)}$ and $H^{(k)}$ are homogeneous forms of degree k, and

$\Phi^{(n)} \neq 0$. Since $H^{(0)}$ is of degree 0 , $H^{(0)} = h$, so

$$H(w,\bar{w}) - h = H^{(1)}(w,\bar{w}) + \dots .$$

From 1.2 and condition 3), Φ and H satisfy the following two relations

a) $$\frac{1}{E}\Phi_{\bar{w}} = H_w$$

b) $$\lim_{p_i \to p_0} \frac{H-h}{|\Phi|} = \lambda .$$

Substituting the Taylor's expansions of Φ and H in a), by comparing degrees we conclude that $H_w^{(k)} = 0$ for $k < n$. But, since H is real,

$$2H_w^{(k)} = H_u^{(k)} - iH_v^{(k)}$$

and hence $H_u^{(k)} = H_v^{(k)} = 0$. Thus $H^{(k)}$ is a constant. But, since $H^{(k)}$ is a homogeneous form of degree k , $H^{(k)} = 0$, $0 < k < n$. Therefore

$$H(w,\bar{w}) - h = H^{(n)}(w,\bar{w}) + H^{(n+1)}(w,\bar{w}) + \dots .$$

Equation b) can be rewritten

$$\lim_{p_i \to p_0} \frac{H^{(n)} + \dots}{|\Phi^{(n)} + \dots|} = \lambda .$$

If polar coordinates r and θ are introduced at p_0 then

$$H^{(m)} = r^m H^{(m)}(\cos\theta, \sin\theta) = r^m H^{(m)}(\theta)$$

$$\Phi^{(m)} = r^m \Phi^{(m)}(\cos\theta, \sin\theta) = r^m \Phi^{(m)}(\theta)$$

where $H^{(m)}$ and $\Phi^{(m)}$ are homogeneous polynomials of degree m in $\cos\theta$ and $\sin\theta$. Therefore

$$\lim_{p_i \to p_0} \frac{r^n H^{(n)}(\theta) + \dots}{|r^n \Phi^{(n)}(\theta) + \dots|} = \lim_{r \to 0} \frac{H^{(n)}(\theta) + rH^{(n+1)}(\theta) + \dots}{|\Phi^{(n)}(\theta) + r\Phi^{(n+1)}(\theta) + \dots|}$$

$$= \frac{H^{(n)}(\theta)}{|\Phi^{(n)}(\theta)|} = \frac{H^{(n)}(w,\bar{w})}{|\Phi^{(n)}(w,\bar{w})|} = \lambda .$$

In our case, $\lambda = 0$ and therefore $H^{(n)}(w,\bar{w}) = 0$. But by equation a)

$$\Phi^{(n)}(w,\bar{w})_{\bar{w}} = H^{(n)}(w,\bar{w})_w = 0 .$$

Therefore $\Phi^{(n)}(w,\bar{w})$ is an analytic function of w . Since it is homogeneous of degree n , $\Phi^{(n)} = cw^n$ where $c \neq 0$. Hence

$$\Phi = cw^n + \Phi^{(n+1)} + \dots$$

$$= cw^n + r^{n+1}B$$

where $|B| < M$ in a sufficiently small neighborhood of $w = 0$.

<u>3.4</u> <u>Second part of the proof</u>: We are now in a position to prove that p_o is in fact isolated and that $j < 0$. Consider the neighborhood $r < \frac{|c|}{M}$. Suppose $\Phi = 0$ at some point $\neq 0$ in this neighborhood. Then

$$cw^n = -r^{n+1}B$$
$$|c|r^n = r^{n+1}|B| \quad .$$

Hence $|c| = r|B|$ for some $r \neq 0, r = \frac{|c|}{|B|} > \frac{|c|}{M}$, which is a contradiction.

To show that $j < 0$, by 1.3 and because $\delta \arg(cw^n) = n\, 2\pi > 0$, it is sufficient to show that

$$\delta(\arg \Phi) = \delta \arg(cw^n)$$

for a small closed curve in the neighborhood $r < \frac{|c|}{M}$. But in this neighborhood

$$|\Phi - cw^n| < |cw^n| \quad .$$

Geometrically, this has the consequence that the three points $\Phi(w)$, 0 , and cw^n never lie on a straight line with 0 between the other two.

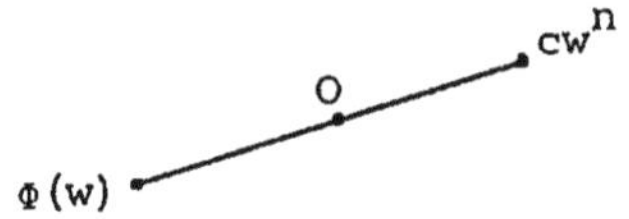

Hence

$$\arg \Phi - \arg cw^n \neq k\pi$$

where k is an odd integer. We have that

$$\frac{\delta(\arg \Phi) - \delta(\arg cw^n)}{2\pi} = \text{integer}$$

while

$$\frac{\arg\Phi - \arg cw^n}{2\pi} \neq \text{half odd integer} \ .$$

Hence by continuity,

$$\delta(\arg \Phi) - \delta(\arg cw^n) = 0 \ .$$

<u>3.5</u> <u>Remarks</u>: The above proof makes strong use of the hypothesis that the surface is analytic, while the proof for surfaces with constant H did not need analyticity. But there is actually no greater generality in the case of constant H . For if the surface is described locally as $z = z(x,y)$ then the equation $H = c$ reads as follows:

$$\varphi(p,q,r,s,t) = (1+q^2)r - 2pqs + (1+p^2)t - 2c(1+p^2+q^2)^{3/2} = 0 \ .$$

This is an elliptic equation since we always have

$$(1+q^2)(1+p^2) - (pq)^2 = 1 + p^2 + q^2 > 0 \ .$$

Hence $\varphi = 0$ is an analytic, elliptic, 2'nd order differential equation. But S. Bernstein's theorem says that a three-times differentiable solution of such an equation is in fact analytic. Hence the surfaces we considered with constant H are in fact analytic.

Hartman and Wintner (Amer. J. of Math., Vol. 76(1954), p. 502) have proved Theorem 3.2 for twice differentiable W-surfaces, using condition 2*) of 3.1. But their proof makes use of the Weingarten relation which ours does not. In fact our proof applies to any analytic surface whose curvature diagram has cusps at the diagonal $(k_1 = k_2)$ with tangents orthogonal to this diagonal.

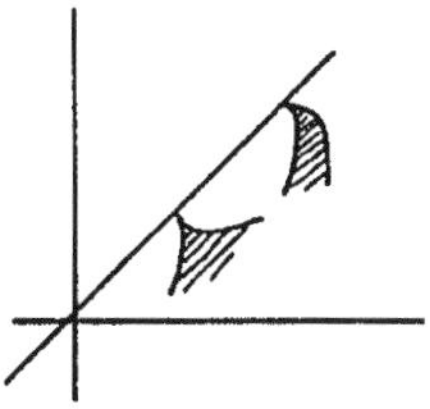

I do not know whether, under these conditions for the diagram, anything can be proved for non-analytic surfaces which are not Weingarten surfaces.

3.6 Further Remarks. For analytic surfaces the following theorems are true. (H. Hopf, Math. Nachrichten, Vol. 4 (1951)).

a) If $\kappa < 0$ (i.e. $|\lambda| < 1$), then $\kappa = -1$ (i.e. $\lambda = 0$) and consequently an umbilic point is isolated and its index is negative.

b) If $\kappa > 0$ (i.e. $|\lambda| > 1$), then $\kappa = (2m+1)^{\pm 1}$ for some positive integer m. An umbilic point is isolated and its index is $+1$.

The reason for these facts is that there are very few pairs of homogeneous forms $\Phi^{(n)}(w,\bar{w})$ and $H^{(n)}(w,\bar{w})$ with $H^{(n)}$ real which satisfy

$$H^{(n)} = \lambda|\Phi^{(n)}| \ , \ H^{(n)}_w = \Phi^{(n)}_{\bar{w}} \ .$$

Although in the quoted paper only Weingarten surfaces are mentioned, the proofs do not use conditions 1) and 2) but only 3). Therefore, the hypothesis that the curvature diagram is a curve can be replaced by the weaker hypothesis that, in the points where $k_1 = k_2$, it has cusps, similar to the figure in 3.5 (but with different tangents).

Recently K. Voss has proved the following theorem for analytic surfaces:

b') If $\kappa > 0$, then the surface is a surface of revolution, with the considered umbilic point on the axis. *)

Exercise I: Show that b) is a corollary of b').

II: Show that the statement about κ in b) breaks down if the surface is not analytic but of class C^n with a given n (which may be large).

Hint: construct a surface of revolution.

---

*) Math. Annalen, Vol. 138 (1959)

CHAPTER VII

Simple Closed Surfaces (of Arbitrary Genus) with Constant Mean Curvature - Generalizations

1. Introduction

In this chapter we will prove that the only simple closed surfaces with constant mean curvature H are the spheres. From this theorem and the main theorem of the preceding chapter it follows that the only undecided cases are non-simple closed surfaces with genus > 0.

A.D. Alexandrov communicated the theorem and sketched the proof in a lecture given at Zurich in July 1955, but the proof has not yet been published (March 1956). The proof depends on the rather obvious observation (to be discussed in Section 2 below) that the spheres are the only closed surfaces with a plane of symmetry in every direction. The proof then comes in two parts, a "geometric" and an "analytic" part. We prove first that under suitable restrictions, any simple closed surface satisfies certain "symmetry" properties, and second that two solutions of an absolutely elliptic second order partial differential equation which have a specified type of contact at a given point actually coincide in a neighborhood of the contact point. The combination of these two results will give us our theorem.

It is my opinion that this proof by A.D. Alexandrov, and especially the geometric part in Section 3 below, opens important new aspects in differential geometry in the large.

2. Another Characterization of the Spheres

2.1 Definition: A plane P in E^3 is a plane of symmetry for a set $S \subset E^3$ if the Euclidean reflection of S in P maps S onto S.

The direction of a plane is the direction of a normal to the plane, and hence by parallel translation corresponds to a unique pair of antipodal points on the unit sphere Σ. The diagram of directions Σ' of S is the set of points on Σ determined as above by all planes of symmetry of S.

2.2 Lemma. If a simple closed surface S has a plane of symmetry in every direction, then S is a sphere. In fact, if the diagram of directions Σ' has an interior point on Σ, then S is a sphere.

Proof: Suppose the north pole is an interior point of Σ' . Then on each great circle through the north pole there is a small arc such that each point on this arc corresponds to a plane of symmetry of S . Now if P_1 and P_2 are two planes of symmetry and if α is the angle between P_1 and P_2 , then a reflection of S in P_1 followed by a reflection in P_2 corresponds to a rotation of S , through an angle 2α about the intersection of P_1 and P_2 , which by definition leaves S invariant. Let P_1 be the plane corresponding to the north pole on Σ' . Then each point on the above small arc of a fixed great circle corresponds to a rotation leaving S invariant. It is clear that all of these rotations are about the same axis. Thus all small rotations about this axis leave S invariant. But the rotations about a fixed axis leaving S invariant form a group which is clearly generated by the "small" rotations. Hence S is invariant under all rotations about this fixed axis; so every point on this great circle corresponds to a plane of symmetry of S . Since the great circle was arbitrary, we conclude that $\Sigma' = \Sigma$. But it follows from this that S is invariant under all rotations. Let $a \in S$. Then S contains a whole sphere through a and therefore S is a sphere.

3. A "Symmetry" Property of Simple Closed Surfaces

3.1 Definition: Let S be a simple closed surface (of arbitrary genus) in E^3 of class C^2 , and let d be a distinguished direction in E^3. (In sketches we will always take d to be the vertical direction oriented from above to below.) Let $n(x)$ denote the inner normal to S

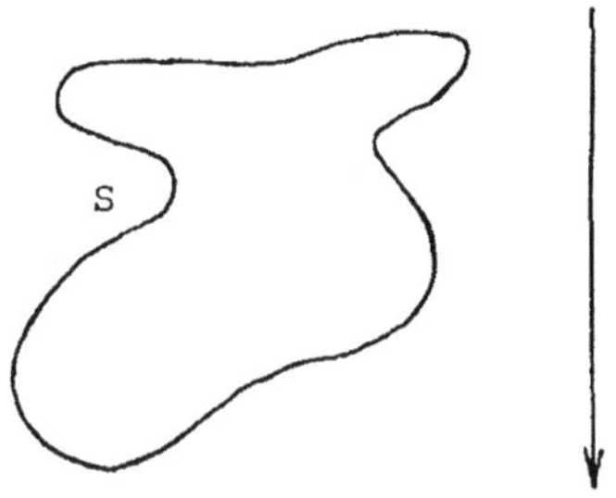

at $x \in S$. Let

$$A = \{x \in S : \sphericalangle[d, n(x)] < \tfrac{\pi}{2}\}$$

$$B = \{x \in S : \sphericalangle[d, n(x)] > \tfrac{\pi}{2}\}$$

$$C = \{x \in S : \sphericalangle[d, n(x)] = \tfrac{\pi}{2}\}$$

I.e., with the above convention, A is the set where the inner normal points downward, B where it points upward, and C where it is horizontal.

A line parallel to d and oriented the same way as d will be called a <u>d-line</u>.

<u>3.2</u> <u>Lemma</u>. 1) A and B are open sets on S and C is closed.

2) If $^{-}$ denotes closure on S, then

$$\bar{A} \subset A \cup C \quad (\bar{A} \cap B \text{ is empty})$$
$$\bar{B} \subset B \cup C \quad (\bar{B} \cap A \text{ is empty}) .$$

3) If ℓ is a straight line in the direction d oriented the same way as d which meets S at $a \in A$, then ℓ is not tangent to S at a, and traversing ℓ in the positive direction, one leaves the exterior of S and enters the interior of S at a. I.e., points of A are points of entrance to the interior of S in the direction d. Similarly, points of B are points of exit from the interior of S in the direction d. The proof is obvious.

<u>3.3</u> <u>Definition</u>: The asymptotic directions at a point of a surface are given by the zeros of the second fundamental form; i.e., by

$$L\,du^2 + 2M\,du\,dv + N\,dv^2 = 0 .$$

At a point where $K > 0$, there are no real solutions and hence no such (real) directions. If $K < 0$ there are exactly two asymptotic directions. We are interested in points p such that $K(p) = 0$. The asymptotic directions at such a point are called <u>doubly asymptotic directions</u>. Two cases are possible: Either $(L,M,N) \neq (0,0,0)$; i.e., p is an ordinary parabolic point, and there is exactly one doubly asymptotic direction, or $(L,M,N) = (0,0,0)$; i.e. p is a flat point, and all tangent directions are doubly asymptotic.

A distinguished direction d in E^3 is called <u>exceptional</u> (with respect to S) if there is a doubly asymptotic direction on S parallel to d.

<u>3.4</u> <u>Lemma</u>. Let S be a simple closed surface of class C^2 and suppose d is a non-exceptional direction with respect to S. Then the set C defined above is the sum of a finite number of non-intersecting simple regular closed curves, and

$$\bar{A} = A \cup C \quad \text{and} \quad \bar{B} = B \cup C .$$

<u>Proof</u>: Let $p \in C$. Let (x,y,z) be a system of rectangular coordinates at p so that the positive x-axis is a d-line, the positive z-axis is the inner normal to S and (x,y,z) is positively oriented. Then the x-y-plane is the tangent plane to S at p and in a neighborhood of

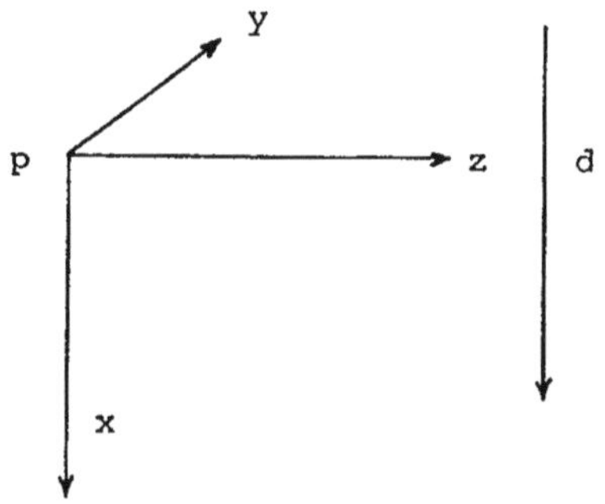

p , the surface is given by $z = z(x,y)$; i.e., if X is the vector describing S , then

$$X = (x,y,z(x,y))$$

where z is twice differentiable. Since

$$X_x = (1,0,z_x)$$
$$X_y = (0,1,z_y)$$

the inner normal is given by

$$\bar{X} = \frac{X_x \times X_y}{|X_x \times X_y|} = \frac{1}{(z_x^2 + z_y^2 + 1)^{1/2}} (-z_x, -z_y, +1) \ .$$

Hence in a neighborhood of p ,

$$A = \{X\colon z_x < 0\} \ , \quad B = \{X\colon z_x > 0\} \ , \quad C = \{X\colon z_x = 0\} \ .$$

Now $z(x,y) = ax^2 + 2\,bxy + cy^2 + D(x,y)$ where D contains terms of higher order. Since p is not a flat point, $(a,b,c) \neq (0,0,0)$. In fact, $(a,b) \neq (0,0)$, for if $(a,b) = (0,0)$, then

$$z(x,y) = cy^2 + D(x,y)$$

which would mean that the x-axis is a double asymptotic direction contrary to the choice of the coordinate system (x,y,z) . Therefore,

$$z_x = 2\,ax + 2\,by + D_x(x,y)$$

where $\operatorname{grad} z_x(0,0) = 2(a,b) \neq (0,0)$. Hence the curve $z_x(x,y) = 0$, which defines C near p , is a regular curve near $(0,0)$. But p

was an arbitrary point of C and hence, since S is compact, C is the union of a finite number of non-intersecting simple regular closed curves.

To prove the second part of the theorem, let $\tilde{C}$ be the projection of C into the x-y-plane. Then $\tilde{C}$ is a regular curve through p such that $z_x > 0$ on one side of $\tilde{C}$ and < 0 on the other side. Hence $C \subset \bar{A} \cap \bar{B}$. Since, by 3.2.2, $\bar{A} \subset A \cup C$ and $\bar{B} \subset B \cup C$, it follows that $\bar{A} = A \cup C$ and $\bar{B} = B \cup C$.

<u>3.5</u> <u>Lemma</u>. With the assumptions and notations of 3.4, assume further that a certain interval $T = \{x: 0 < x < x_1\}$ of the x-axis does not contain any point of S so that T lies either in the interior or the exterior of S. Let $\hat{S}$ be the intersection of S with the half plane $(y = 0,\ x > 0)$ in an arbitrarily small neighborhood of p. Then either T is in the interior of S and $\hat{S} \cap A \neq 0$ or T is in the exterior of S and $\hat{S} \cap B \neq 0$.

<u>Proof</u>: As in 3.4 the positive z-axis lies in the interior of S and the negative z-axis lies in the exterior of S (in a neighborhood of p). If T is in the interior, then S separates T from the negative z-axis (see Figure 1) and therefore $z < 0$ on $\hat{S}$. Consequently, since $z = 0$ at p, $z_x < 0$ somewhere on $\hat{S}$; i.e., $\hat{S} \cap A \neq 0$. Similarly, if T is in the exterior, then S separates T from the positive z-axis and hence $z > 0$ on $\hat{S}$, which implies $z_x > 0$ somewhere on $\hat{S}$; i.e., $\hat{S} \cap B \neq 0$. (see Figure 2)

<u>Fig. 1</u>

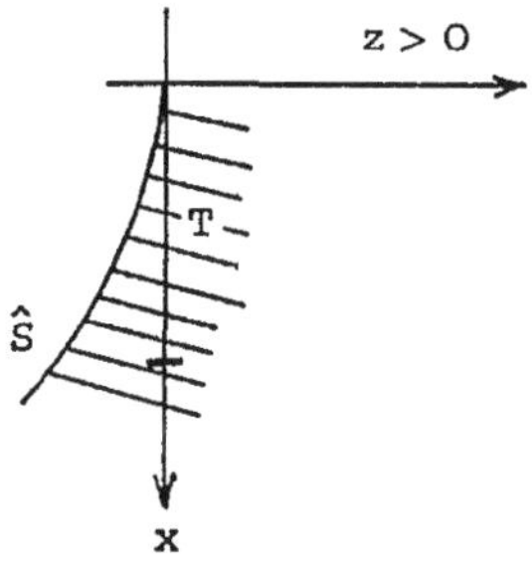

<u>Fig. 2</u>

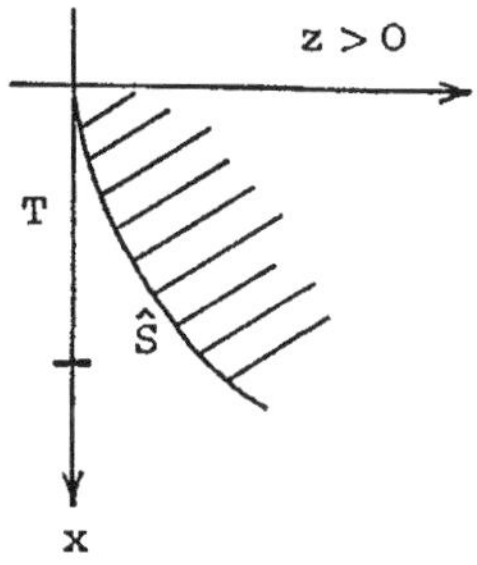

<u>3.6</u> <u>Definition</u>. Let α and β be two oriented pieces of surfaces with interior points in common, and let p be such a common point.

1) If α and β have a common tangent plane at p, then α and β have a <u>contact</u> at p.

2) If α and β have a contact at p such that the positive normals to α and β coincide at p, then α and β have a <u>posi-</u>

tive contact at p.

3) In a neighborhood of p, let α be given by $z_1 = z_1(x,y)$ and β by $z_2 = z_2(x,y)$. If $z_1 - z_2$ changes sign in each neighborhood of p, then α and β intersect at p.

4) If α and β have a contact at p but do not intersect at p, then α and β have a proper contact at p.

5) If α and β have a contact at p such that there is on α a regular (i.e., of class C^1) curve Γ through p with the property that on at least one side of Γ, α and β do not intersect, then α and β have a semi-proper contact at p.

3.7 Example. The surfaces in E^3 defined by the equations

$$z^k = \mathrm{Im}\{c(x+iy)^k\}, \quad k \geq 2$$

have a contact with the x-y-plane at the origin which is not semi-proper.

Proof: The function z^k is zero on each line through the origin which makes an angle $j\frac{\pi}{k}$ with the x-axis, $j = 1,\dots,k$, and changes sign on crossing any of these lines. If Γ is any regular curve through the origin, since $k \geq 2$, a zero line of z^k points into each component of the complement of Γ in a neighborhood of the origin. Since z^k changes sign across these lines, it changes sign in each such component

3.8 Theorem. Let S be a simple closed surface of class C^2 and suppose d is a non-exceptional direction with respect to S. Then there is a plane P perpendicular to d such that if S' is the reflection of S in P, then S and S' have a positive semi-proper contact, (the "positive" normals always being the interior normals).

Proof: If a and b are two points, let $M(a,b)$ denote the plane which is the perpendicular bisector of the line joining a and b. We will prove that either

1) There are points $a \in A$ and $b \in B$ which are on a d-line such that if β is a sufficiently small neighborhood of b on S and β' is the reflection of β in $M(a,b)$, then β' and A do not intersect; or

2) There is a point $c \in C$ such that if γ is a sufficiently small neighborhood of c on S and γ' is the reflected image of γ in the plane through c perpendicular to d, then γ' does not have an intersection with A.

Assume again that d is the vertical direction and let P be a horizontal plane below S. Let $\bar{B}''$ be the reflection of $\bar{B}$ in P. Translate $\bar{B}''$ upwards until it first meets $\bar{A}$. Call this translated set $\bar{B}'$. Then $\bar{B}'$ is the reflection of $\bar{B}$ in a plane parallel to P and clearly $\bar{B}'$ and $\bar{A}$ have a common point, but no intersection. Let $p \in \bar{A} \cap \bar{B}'$. Then we will show that either $p = a \in A$ and 1) above is satisfied or $p = c \in C$ and 2) is satisfied.

We observe first that there is no point of $\bar{A} = A \cup C$ below p on the d line through p since $\bar{B}''$ is translated upwards until it first meets $\bar{A}$. Hence if there is any point of S below p, it is a point of B. But, by 3.2.3, such a point is an exit point from the interior of S, and since there are no points of $A \cup C$ below p, it is the only point of S below p. Thus

1) either there is exactly one point b of B but no point of $A \cup C$ below p,

2) or there is no point of S below p.

<u>Case 1</u>. We prove that in this case $p \in A$ (i.e., $p \notin C$). Since $b \in B$, the d line through p is not tangent to S at b and therefore is also not tangent to B' at p. (see the figure) From the definition of B', it follows that in a neighborhood of p, there is no point of A below B'. On the other hand, since b is a point of exit from the interior of S the segment $\overline{pb}$ lies in the interior. Therefore, if p were a point of C, it would follow from Lemma 3.5 that in each neighborhood of p, there are points of $\hat{S} \cap A$ where the curve $\hat{S}$ is tangent to the x-axis and below the plane $x = 0$, which is obviously a contradiction. Consequently, $p \in A$.

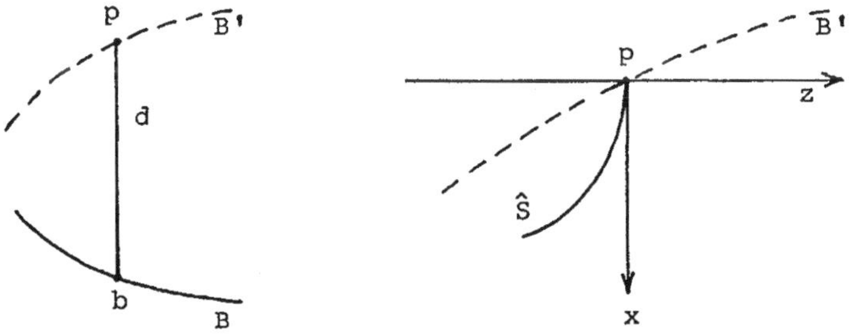

<u>Case 2</u>. In this case $p \in C$, for suppose $p \in A$. Then the x-axis enters the interior of S at p and hence must leave the interior at some

point $b \in B$ contrary to the assumption that there is no point of S below p. Therefore $p = c \in \bar{A} \cap \bar{B}$. Under the map of $\bar{B}$ into $\bar{B}'$, $p = c$ clearly is mapped into itself. Hence the mapping is a reflection in a horizontal plane through p.

Let γ be a neighborhood of p, and γ' the reflected neighborhood. Now γ' contains points of A', B', and C' (where $'$ denotes the reflected sets). We know already that $\bar{B}' = B' \cup C'$ does not intersect A, so it remains to show that $\gamma' \cap A'$ does not intersect A.

Let $\tilde{C}$ be the projection of C into the tangent x-y-plane. Then by 3.4, $\tilde{C}$ is a regular curve. Let $q \in C$ and let q' be the reflected point of q. Then q' is either below q or equal to q, since p is the first contact point of $\bar{B}''$ and $\bar{A}$. Thus

$$x(q) \leq x(q') = -x(q)$$

and hence $x(q) \leq 0$. Therefore $\tilde{C}$ is above or on the y-axis and has a minimum at p.

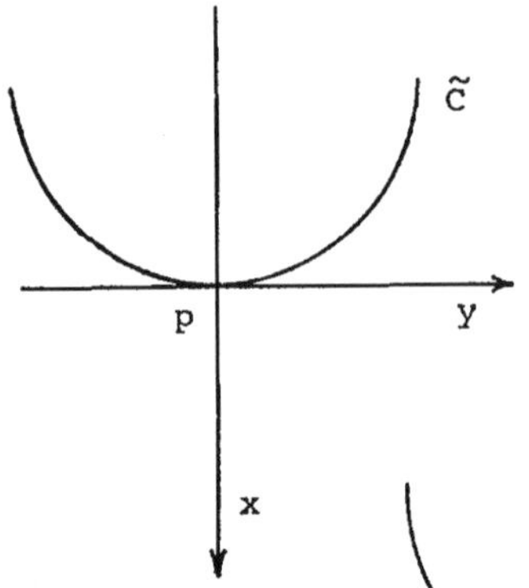

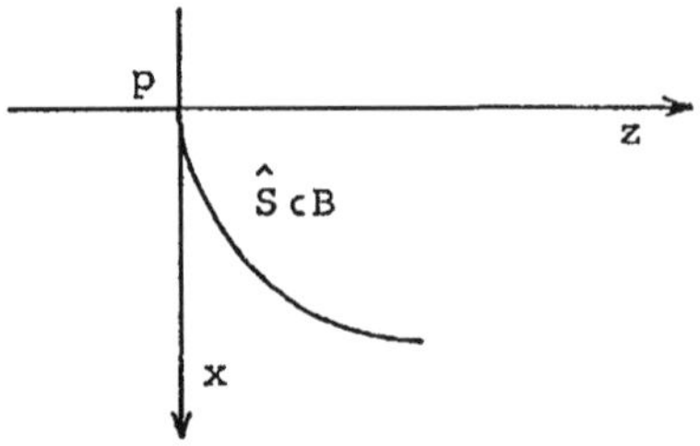

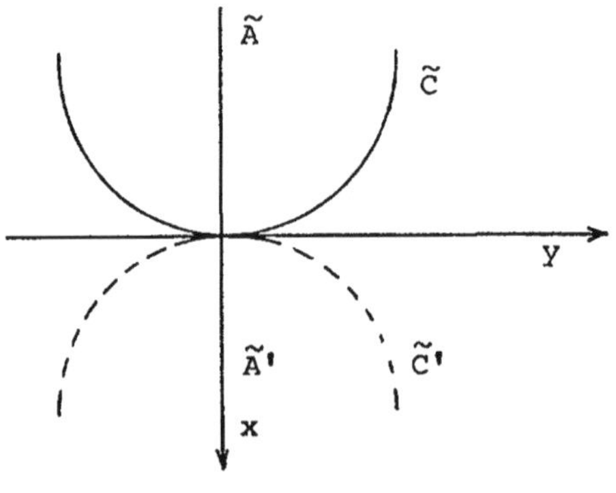

$\tilde{C}$ decomposes the neighborhood of p into two components, one of them corresponding to the projection $\tilde{A}$ of A, and the other to the projection $\tilde{B}$ of B. We claim that $\tilde{B}$ is the lower one; i.e., the one containing the positive x-axis. Indeed, the positive x-axis lies in the exterior of S and therefore according to lemma 3.5, there are points of B below the y-z-plane. It follows that $\tilde{A}$ is above $\tilde{C}$.

But then it is obvious from the drawing that $\tilde{A}'$ and $\tilde{C}'$ have no points in common with $\tilde{A}$ and therefore $\gamma' \cap A'$ does not intersect A.

This completes the proof of Theorem 3.8.

<u>3.9</u> <u>Examples</u>: Both cases of Theorem 3.8 actually occur, as is illustrated by the following two examples.

<u>Case 1</u>

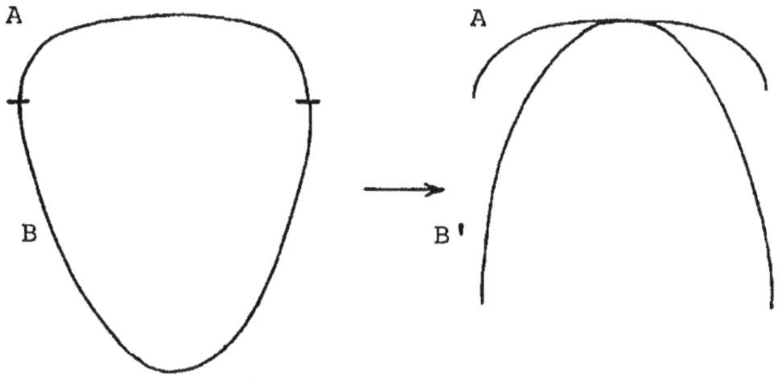

<u>Case 2</u>

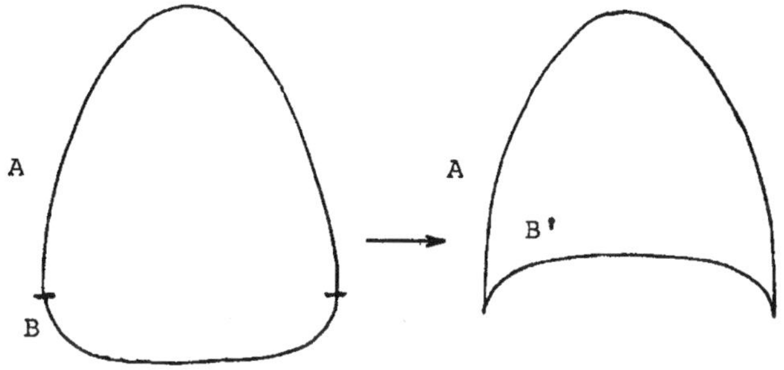

The proof fails for surfaces with self-intersections, as illustrated below.

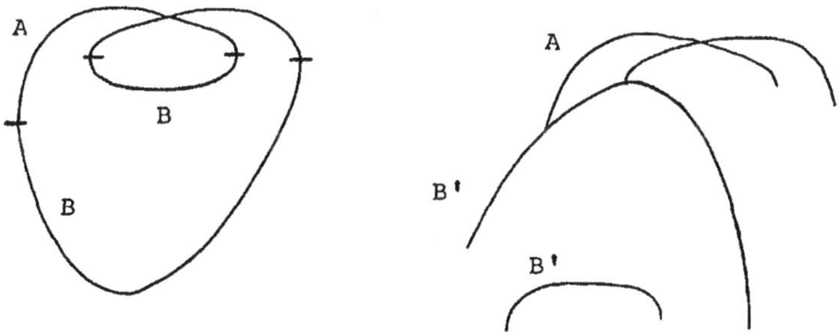

However, the theorem may very well be true, even though the proof does fail.

4. Absolutely Elliptic Partial Differential Equations

4.1 Definition. Let $\varphi = \varphi(r,s,t,p,q,z,x,y)$ be a function of eight variables defined in some region in E^8. Then if φ has continuous first partial derivatives we may regard $\varphi = 0$ as a second order partial differential equation for $z = z(x,y)$ with partial derivatives, $p = z_x$, $q = z_y$, $r = z_{xx}$, $s = z_{xy}$, and $t = z_{yy}$.

Consider the quadratic form

$$\Delta = \varphi_r \lambda^2 + \varphi_s \lambda\mu + \varphi_t \mu^2 .$$

1) $z(x,y)$ is called an elliptic solution of $\varphi = 0$, or $\varphi = 0$ is said to be elliptic with respect to z if Δ is positive definite when the particular function $z(x,y)$ is substituted in φ.

2) $\varphi = 0$ is elliptic if Δ is positive definite for every solution $z(x,y)$.

3) $\varphi = 0$ is absolutely elliptic in a region $R \subset E^8$ if Δ is positive definite for arbitrary values of its eight arguments in R.

If $\varphi = A_1 r + A_2 s + \ldots + A_6 z + A_7$ where $A_i = A_i(x,y)$, $i = 1,\ldots,7$, then $\varphi = 0$ is a linear partial differential equation. If $A_7 = 0$, then φ is homogeneous. A linear equation is elliptic if $A_1\lambda^2 + A_2\lambda\mu + A_3\mu^2$ is positive definite.

4.2 Example. The equation $H = c$ is

$$(1+q^2)r - 2pqs + (1+p^2)t - 2c(1+p^2+q^2)^{3/2} = 0 .$$

Hence $\varphi_r = 1+q^2$, $\varphi_s = -2pq$, $\varphi_t = 1+p^2$. Therefore,

$$\begin{aligned}\Delta &= (1+q^2)\lambda^2 - 2pq\,\lambda\mu + (1+p^2)\mu^2 \\ &= \lambda^2 + \mu^2 + (q\lambda - p\mu)^2\end{aligned}$$

which is clearly positive definite for every value of the arguments. Thus $H = c$ is an absolutely elliptic partial differential equation.

4.3 Lemma. Let $\psi(u)$ be a function of n variables, $u = (u_1,\ldots,u_n)$ which is differentiable in a convex region in E^n. Then

$$\psi(v) - \psi(u) = \sum_{i=1}^{n} A_i(v_i - u_i)$$

where

$$A_i(u,v) = \int_0^1 \psi_i(\tau v + (1-\tau)u)\,d\tau$$

and ψ_i is the derivative of ψ with respect to its i'th argument.

Proof: Let $'$ denote differentiation with respect to the parameter τ. Then

$$\psi(v) - \psi(u) = \int_0^1 \psi'(\tau v + (1-\tau)u)\,d\tau$$

$$= \int_0^1 \sum_{i=1}^n \psi_i(\tau v + (1-\tau)u)(\tau v_i + (1-\tau)u_i)'\,d\tau$$

$$= \sum_{i=1}^n \left\{\int_0^1 \psi_i(\tau v + (1-\tau)u)\,d\tau\right\}(v_i - u_i)$$

4.4 Lemma. Let $\varphi = 0$ be a partial differential equation which is absolutely elliptic in a convex region R and let z_1 and z_2 be two solutions of $\varphi = 0$. Then

$$Z = z_2 - z_1$$

satisfies a linear homogeneous elliptic partial differential equation.

Proof: By assumption

$$\varphi(r_i, s_i, t_i, p_i, q_i, z_i, x, y) = 0, \quad i = 1,2 \quad .$$

Hence, by Lemma 4.3,

$$\varphi(r_2, s_2, \ldots) - \varphi(r_1, s_1, \ldots)$$

$$= A(r_2 - r_1) + B(s_2 - s_1) + C(t_2 - t_1) + D(p_2 - p_1) + E(q_2 - q_1)$$
$$+ F(z_2 - z_1) = 0$$

where $A(x,y) = \int_0^1 \varphi_r\,d\tau$ and the arguments of φ_r are $\tau r_2(x,y) + (1-\tau)r_1(x,y), \ldots, \tau z_2(x,y) + (1-\tau)z_1(x,y)$, x,y. It is clear that this equation is homogeneous and linear, and that Z satisfies this equation.

The equation is elliptic since

$$\varphi_r\lambda^2 + \varphi_s\lambda\mu + \varphi_t\mu^2$$

is positive definite for arbitrary values in R, and hence in particular for the values which appear in the integrals for A, B, etc. Therefore integrating the form, we get

$$\int_0^1 [\varphi_r\lambda^2 + \varphi_s\lambda\mu + \varphi_t\mu^2]\,d\tau = A(x,y)\lambda^2 + B(x,y)\lambda\mu + C(x,y)\mu^2$$

which is also positive definite.

4.5 Theorem. Let $\varphi = 0$ be a partial differential equation which is absolutely elliptic in a convex region. Let z_1 and z_2 be two solutions of $\varphi = 0$ such that at $(0,0)$

$$z_1(0,0) = z_2(0,0)$$
$$p_1(0,0) = p_2(0,0)$$
$$q_1(0,0) = q_2(0,0)$$

but $z_1 \not\equiv z_2$ in a neighborhood of $(0,0)$. Then the surface defined by

$$Z = z_2 - z_1$$

has a contact with the x-y-plane which is not semi-proper.

We will prove this under the assumption that Z is analytic. This is the case, for example, if φ is analytic and Z is at least three times differentiable, by Bernstein's Theorem.

Proof: By Lemma 4.4, Z satisfies a linear, homogeneous elliptic partial differential equation.

$$\Sigma A_{ij} \frac{\partial^2 Z}{\partial x_i \partial x_j} + \Sigma B_i \frac{\partial Z}{\partial x_i} + FZ = 0 .$$

Under a homogeneous affine transformation of coordinates, the coefficients A_{ij} behave like the coefficients of a quadratic form. I.e., if $u_k = \Sigma t_{ki} x_i$ where t_{ki} are constants and $\det(t_{ki}) \neq 0$ then $\Sigma A_{ij} \frac{\partial^2 Z}{\partial x_i \partial x_j}$ transforms to $\Sigma \bar{A}_{ij} \frac{\partial^2 Z}{\partial u_i \partial u_j}$, where

$$(\bar{A}_{ij}) = (t_{ij})(A_{ij})(t_{ij})' .$$

Since the form given by the A_{ij} is positive, we may change coordinates so that

$$A_{11}(0,0) = 1 , \quad A_{12}(0,0) = 0 , \quad A_{22}(0,0) = 1 .$$

Now Z is assumed to be analytic and hence can be expanded in homogeneous forms,

$$Z = Z^{(n)}(x,y) + Z^{(n+1)}(x,y) + \ldots$$

where $Z^{(n)} \not\equiv 0$, $n \geq 2$. If this expression is substituted in the above transformed equation it is easy to see that the terms of lowest order have order $n-2$ and that these terms come from $Z^{(n)}_{xx}$ and $Z^{(n)}_{yy}$. Since $A_{11}(0,0) = A_{22}(0,0) = 1$, we have

$$\Delta Z^{(n)} = Z^{(n)}_{xx} + Z^{(n)}_{yy} = 0$$

where Δ is the Laplace operator. But the only forms satisfying this are the forms

$$Z^{(n)} = \mathrm{Im}\{c(x+iy)^n\} .$$

We have already seen in Example 3.7 that these forms have the desired property. But on each ray through $(0,0)$ on which $Z^{(n)} \neq 0$ there is an interval containing $(0,0)$ in which Z has the same sign as $Z^{(n)}$. Hence Z has the desired property.

4.6 Corollary: Two regions of surfaces satisfying the same absolutely elliptic partial differential equation and which have a positive semi-proper contact are identical in a neighborhood of the contact.

4.7 A Special Case. (not needed for our main theorem): Consider two pieces of surfaces with the same constant Gauss curvature c. They satisfy the partial differential equation

$$\varphi = rt - s^2 - c(1+p^2+q^2)^2 = 0 .$$

So

$$\varphi_r\varphi_t - \frac{1}{4}\varphi_s^2 = tr - s^2 = c(1+p^2+q^2)^2 .$$

Hence this equation is elliptic if $c > 0$.

Now let $c > 0$. Then the only place where trouble can occur in the application of 4.6 is in the convexity of the domain of ellipticity of this equation. What we really need is the convexity with respect to r,s,t. Hence we are concerned with the part of (r,s,t)-space where $rt - s^2 > 0$. Now in E^3, the locus of the equation $rt-s^2 = 0$ is a cone. This is easily seen by making the change of coordinates $r = \xi + \eta$, $t = \xi - \eta$, $s = \zeta$. The equation then becomes $\xi^2 - \eta^2 - \zeta^2 = 0$. If the left hand side is positive, then $\xi \neq 0$. Hence we are concerned with the region where $(\eta/\xi)^2 + (\zeta/\xi)^2 < 1$. This is the interior of the cone $(\eta/\xi)^2 + (\zeta/\xi)^2 = 1$. Hence our region is the union of two sets, each of which is convex.

In one set, $r > 0$ and hence $t > 0$ and in the other $r < 0$ and hence $t < 0$. Thus if z_1 and z_2 are two solutions of $K = c$ such that r_1 and r_2 have the same sign, then $z_1 - z_2$ satisfies a linear homogeneous elliptic partial differential equation.

Therefore, it follows that two pieces of surfaces with the same constant positive Gauss curvature which have a contact such that both of them are on the same side of the tangent plane are identical.

5. The Main Theorem

5.1 Lemma. Let S be a simple closed surface of class C^3 with constant mean curvature. Then the set of non-exceptional directions, considered as a set on the unit sphere Σ, has an interior point.

Proof: Since $H = c$ is an analytic equation, by Bernstein's Theorem, S is analytic. From this we can prove that the set of non-exceptional directions is all of Σ except perhaps for an analytic curve on Σ.

1) There are no flat points on S. For suppose p is a flat point. Then $k_1(p) = k_2(p) = 0$. Hence $k_1(p) + k_2(p) = 2c = 0$, so $c = 0$. But then $k_2 = -k_1$ and $K = -k_1^2$ which is always non-positive, which contradicts II, 4.2.

2) The set of parabolic points is the analytic curve defined by $K = 0$. In each of these points there is exactly one double asymptotic direction. It is clear that these directions describe an analytic curve on the sphere Σ of all directions.

5.2 Theorem. Let S be a simple closed surface of class C^3 with constant mean curvature. Then S is a sphere.

Proof: If d is a non-exceptional direction, then by Theorem 3.8, there is a plane P such that if S' is the reflection of S in P, then S' and S have a positive semi-proper contact. Hence by 4.6, and since $H = c$ is an absolutely elliptic equation (see 4.2), S and S' coincide in a neighborhood of the contact. But if two analytic surfaces coincide in a neighborhood then they are indentical. Therefore $S' = S$. Hence the non-exceptional directions correspond to directions of planes of symmetry. Consequently, by Lemma 5.1, the set of directions of planes of symmetry has a non-empty interior. Therefore, by Lemma 2.2, S is a sphere.

6. Generalizations - Simple Closed Weingarten Surfaces

6.1 Lemma. Let S be a closed Weingarten surface whose Weingarten relation $U(K,H) = 0$ corresponds to a partial differential equation $\varphi = 0$ which is elliptic for S. Then there are no flat points on S.

Proof: We saw in V, 2.4 that the equation of a Weingarten surface is elliptic if and only if $\frac{dk_2}{dk_1} < 0$. Now a flat point corresponds to a point at the origin on the curvature diagram of S. But if $\frac{dk_2}{dk_1} < 0$, then the existence of such a point implies that the entire curvature

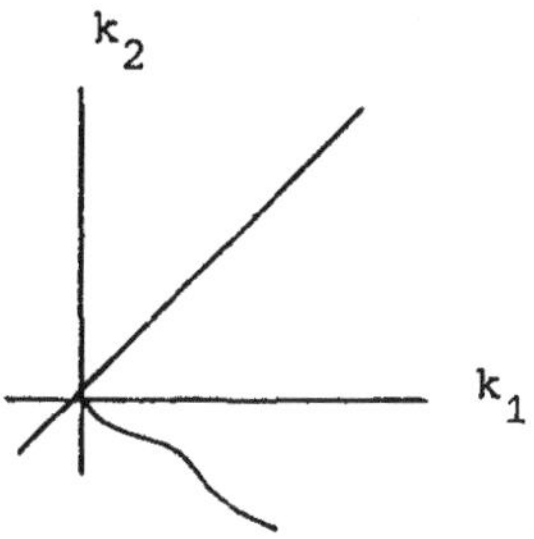

diagram lies in the 4'th quadrant which in turn implies that $K = k_1 k_2 \leq 0$, which contradicts II, 4.2.

6.2 Theorem. Let S be a simple closed W-surface of class C^3 whose Weingarten relation $U(K,H) = 0$ corresponds to a partial differential quation $\varphi = 0$ which is analytic and absolutely elliptic. Then S is a sphere. The proof goes exactly as in Section 5.

6.3 Possible Generalizations. The proof of Theorem 6.2 depended both on the analyticity and the absolute ellipticity of φ . In Theorem 4.5, the requirement that φ be analytic can be removed entirely. It is in fact sufficient to assume that φ is of class C^2 . For details, see the articles of E. Hopf in the Proceeding of the Academy in Berlin, 1927, and the Proceedings of the A.M.S., 1952. We also used analyticity in 5.1 and 5.2, but it seems very likely that this can also be easily avoided.

6.4 Remarks on Possible Generalizations. It is also not necessary for φ to be absolutely elliptic. For suppose φ is only elliptic for S . Then S satisfies another equation which is absolutely elliptic. For, since φ is elliptic for S , the diagram of S is a monotone decreasing curve, as illustrated.

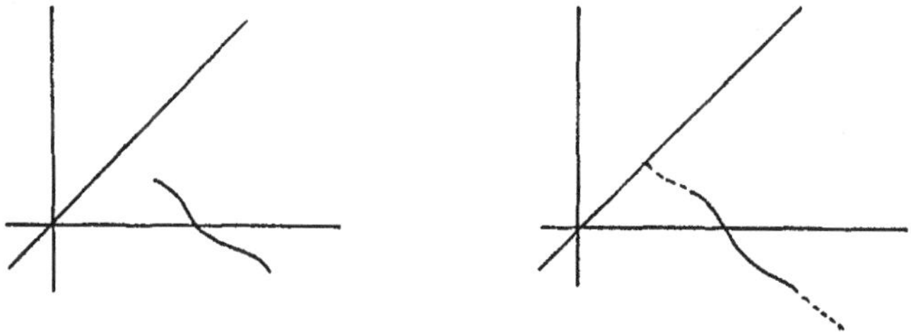

Extend this curve by a C^n curve to the diagonal line and to all values > 0 of k_1 such that the resulting curve is always monotone decreasing and such that $\frac{dk_2}{dk_1} = -1$ at the diagonal line. The lines $k_1 + k_2 = \text{const.}$ and $k_1 - k_2 = \text{const.}$ define a new coordinate system. It is clear that in this coordinate system, the curve can be written

$$k_1 + k_2 = f(k_1 - k_2)$$

or

$$H = f(H^2 - K)$$

where the function f is defined for all non-negative values of its arguments.

Now consider the Weingarten relation

$$U^*(K,H) = H - f(H^2 - K) = 0 .$$

Then the corresponding equation $\varphi^* = 0$ is absolutely elliptic in the closed half-plane below the main diagonal, and clearly S satisfies this equation.

It is very likely that if one carries through all these details, one gets a proof of the fact that A. Alexandrov's Theorem 6.2 holds for all simple closed surfaces of class C^3 which fulfill a Weingarten relation $U(K,H) = 0$, where U is differentiable and

$$U_{k_1} U_{k_2} > 0 .$$

CHAPTER VIII

The Congruence Theorem for Ovaloids

1. The Second Fundamental Forms of Isometric Surfaces

1.1 Introduction. Let S and S^* be two isometric surfaces (see I, 2.6). Let h be the isometry between S and S^* and let u, v be parameters such that $X(u,v)$ and $X^*(u,v)$ are corresponding points under the map h. Then S and S^* have the same first fundamental forms; i.e.

1) $$(E,F,G) = (E^*,F^*,G^*) .$$

We wish to prove that if S and S^* are isometric ovaloids, then S and S^* are congruent. We will prove, in fact, that if h is an isometry, then h is a (proper or improper) Euclidean motion. Theorem 1.2 below will show that it is sufficient to prove that S and S^* have the same second fundamental forms.

Since S and S^* are isometric surfaces, they have a common parameter surface S_o (in the case of ovaloids we may assume that S_o is a sphere). Hence we may regard the forms

2) $$L\,du^2 + 2M\,dudv + N\,dv^2$$
$$L^*du^2 + 2M^*dudv + N^*dv^2$$

as being forms on the same surface, S_o .

In the case $K>0$, both forms are definite, and we may assume they are both positive definite, since they can be made positive by a reflection.

Then we wish to show that

3) $$(L,M,N) = (L^*,M^*,N^*) .$$

Or, equivalently, if
$$\lambda = L^*- L,\ \mu = M^*- M\ ,\ \nu = N^*- N$$
then

(3') $$(\lambda,\mu,\nu) = (0,0,0) \quad \text{on } S_o .$$

Since S and S^* are isometric, $K = K^*$ and hence
$$\frac{LN-M^2}{EG-F^2} = \frac{L^*N^*-M^{*2}}{E^*G^*-F^{*2}} .$$

Consequently, by 1)

4) $$LN- M^2 = L^*N^*- M^{*2} .$$

1.2 Theorem. Let S and S^* be two surfaces such that there is a 1-1 correspondence between them under which they have the same first and second fundamental forms. Then S and S^* are congruent.

Proof: In I, 9.1 we recalled that the theorem is true in the small; i.e., if a and a^* are corresponding points of S and S^*, then there are neighborhoods A and A^* which are congruent under a Euclidean motion M_a. We wish to show that if b and b^* are any other pair of corresponding points then $M_b = M_a$.

Now A and B can be joined by a finite chain of neighborhoods satisfying the above properties. Hence it is sufficient to consider neighborhoods A and B which have a non-empty intersection C. But on C, M_a and M_b are both given by the isometry h, and hence they are identical on C. Therefore, since M_a and M_b are Euclidean motions which agree on an open set of a surface, it follows that $M_a = M_b$.

1.3 Theorem. Let

$$L\,du^2 + 2M\,du\,dv + N\,dv^2$$
$$L^*du^2 + 2M^*du\,dv + N^*dv^2$$

be two positive quadratic forms such that $LN - M^2 = L^*N^* - M^{*2}$. Let $\lambda = L^* - L$, $\mu = M^* - M$, $\nu = N^* - N$. Then the form

$$\lambda du^2 + 2\mu du\,dv + \nu\,dv^2$$

is either indefinite or identically zero; i.e. $\lambda\nu - \mu^2 \leq 0$ and $= 0$ if and only if $\lambda = \mu = \nu = 0$.

Proof: The equality $LN - M^2 = L^*N^* - M^{*2}$ remains valid after a linear transformation of coordinates. But since the forms are positive definite, we can transform both of them simultaneously to canonical form. Thus we may assume that $M = M^* = 0$ and hence $\mu = 0$. Then $L^*N^* = LN$ where all four terms are necessarily positive. Now either $L^* = L$ or $L^* \neq L$. If $L^* = L$, then $N^* = N$ and $\lambda = \nu = 0$ and hence the form is identically zero. If $L^* \neq L$, we may assume $L^* > L$ and hence $N^* < N$. But then $\lambda > 0$, $\nu < 0$ and $\mu = 0$ which implies that the form is indefinite.

Remark. In order to prove our main theorem it is therefore sufficient to prove that $\lambda\nu - \mu^2 = 0$.

1.4 Theorem. The functions λ, μ and ν satisfy

$$-\lambda\nu + \mu^2 = N\lambda - 2M\mu + L\nu .$$

Proof:
$$\begin{vmatrix} L^* & M^* \\ M^* & N^* \end{vmatrix} = \begin{vmatrix} L & M \\ M & N \end{vmatrix}$$

where $L^* = L + \lambda$, $M^* = M + \mu$, $N^* = N + \nu$. Hence
$$\begin{vmatrix} L & M \\ M & N \end{vmatrix} + \begin{vmatrix} \lambda & \mu \\ \mu & \nu \end{vmatrix} + (L\nu - 2M\mu + N\lambda) = \begin{vmatrix} L & M \\ M & N \end{vmatrix} .$$

Therefore $-\lambda\nu + \mu^2 = N\lambda - 2M\mu + L\nu$.

Corollary. If $K > 0$, then there exists a positive quadratic form (L,M,N) such that
$$-\lambda\nu + \mu^2 = N\lambda - 2M\mu + L\nu .$$

1.5 Definition. Let $J(f,g,h)$ denote the ideal generated by the continuous functions f, g, and h in the ring of all continuous functions. That is, $J(f,g,h)$ is the set of all continuous functions of the form
$$af + bg + ch$$
where a, b, and c are continuous functions. We will be interested in functions which are zero mod J.

1.6 Theorem.
$$\lambda_v - \mu_u = 0 \bmod J(\lambda,\mu,\nu)$$
$$\mu_v - \nu_u = 0 \bmod J(\lambda,\mu,\nu)$$

(Functions λ,μ,ν satisfying this set of equations are called "pseudo-Codazzi" functions.)

Proof: The Codazzi equations for L, M, and N are
$$L_v - M_u = a_1 L + a_2 M + a_3 N$$
$$M_v - N_u = b_1 L + b_2 M + b_3 N$$
where the a's and b's are given in terms of the first fundamental form. Hence L^*, M^*, and N^* satisfy the same equations. Thus, by subtracting the two pairs of equations, we get
$$\lambda_v - \mu_u = a_1\lambda + a_2\mu + a_3\nu$$
$$\mu_v - \nu_u = b_1\lambda + b_2\mu + b_3\nu .$$

Remark. This theorem is non-trivial only at the common zeros of λ, μ and ν. For, suppose $\lambda \neq 0$. Then
$$\lambda_v - \mu_u = a\,\lambda$$
where $a = \dfrac{\lambda_v - \mu_\nu}{\lambda}$.

But the theorem tells us for instance that where λ , μ , and ν have a common zero, then also

$$\lambda_v - \mu_u = 0$$
$$\mu_v - \nu_u = 0 .$$

2. Nets of Curves and their Singularities

2.1 Discussion. Let $A\,du^2 + 2B\,dudv + C\,dv^2$ be a quadratic form with $AC - B^2 < 0$. Then in the small this determines two families of curves which form a net in the small.

The question then arises; suppose that in a region of a surface we have such a net in the small in a neighborhood of every point, does this imply that we have a net in the large formed by two families of curves which can be distinguished from each other? The situation illustrated below shows that in general this is not true. This family of

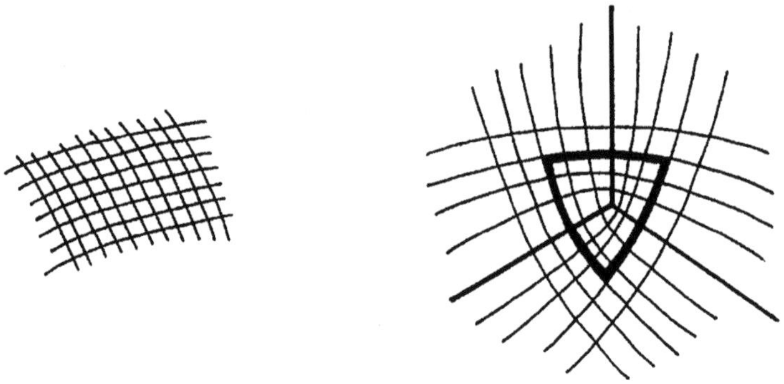

lines has the property that if a point on the heavy triangle is assigned a line element determined by the family of curves and if this line element is extended continuously around the triangle, then in going around the triangle we come back to an element belonging to the other line element in the point in which we started. Hence this set of curves cannot be decomposed into two families of curves. We will show that this does not happen in the case in which we are interested.

2.2 Theorem. Let $A\,du^2 + 2B\,dudv + C\,dv^2$ be a quadratic form on a sphere such that either $AC - B^2 < 0$ or $A = B = C = 0$. Let G be a region in which we do not have $A = B = C = 0$. Then the equation

$$A\,du^2 + 2B\,dudv + C\,dv^2 = 0$$

determines exactly two families of curves in G .

Proof: Since we are on a sphere, and since the neglection of a single point (the point ∞) does not affect the discussion, we can use a single coordinate system throughout the proof. If the equation is multiplied by A, then it can be written

$$(A\,du + B\,dv)^2 + (AC - B^2)dv^2 = 0 .$$

Let $-D^2 = AC - B^2$ where $D > 0$. Then the equation factors into the two equations

I) $A\,du + (B+D)dv = 0$ II) $A\,du + (B-D)dv = 0$.

This determines two families of curves unless one of the equations happens to be identically zero. This happens in Case I, for example, if $A = 0$ and $B < 0$, for then $D = -B$.

On the other hand, if we had multiplied the original equation by C first, and factored we would have gotten the equivalent equations

I') $(B-D)du + C\,dv = 0$ II') $(B+D)du + C\,dv = 0$.

Hence we have two well-defined families of curves given by the following pairs of equations:

I) $A\,du + (B+D)dv = 0$, $(B-D)du + C\,dv = 0$ II) $A\,du + (B-D)dv = 0$, $(B+D)du + C\,dv = 0$.

<u>Exercise</u>: Show that the above theorem is true for any orientable surface.

<u>2.3</u> <u>Theorem</u>. Using the notation of Theorem 2.2, let p be an isolated singularity of

$$A\,du^2 + 2B\,dudv + C\,dv^2 = 0$$

and let j_1 and j_2 be the indices of p with respect to the curves I and II respectively. Then $j_1 = j_2 = j$, where

$$j = -\frac{1}{2}\frac{1}{2\pi}\delta_c\ [\arg(A-iB)] .$$

<u>Proof</u>: It is clear that $j_1 = j_2$ since the two families of curves are distinct at every point, and since the angle between them is never a multiple of π. We set $j = j_1$.

Now consider the family of nets

$$A\,du^2 + 2B\,dudv + [(1-t)C - tA]dv^2 = 0 , \quad 0 \leq t \leq 1 .$$

The discriminant is given by

$$\begin{aligned}&(1-t)AC - (1-t)B^2 - tA^2 - tB^2\\ &\quad = (1-t)(AC-B^2) - t(A^2+B^2) < 0 , \quad 0 \leq t \leq 1 .\end{aligned}$$

Hence by Theorem 2.2 we get a net of curves for each t, all with the same isolated singularity p . Hence the index j_t is well defined for every t and, by continuity, $j = j_t$, $0 \leq t \leq 1$. Therefore j is the same as the index for the field

$$A\,du^2 + 2B\,du dv - A\,dv^2 = 0 .$$

Or
$$A(du^2 - dv^2) + 2B\,du dv = 0 .$$

Choose one family of curves and let τ be the angle between this family and the u-direction. Then

$$du : dv = \cos\tau : \sin\tau$$

and
$$A\cos 2\tau + B\sin 2\tau = 0 .$$

Now if $A = B = 0$, then $AC - B^2 = 0$ and hence $C = 0$, which is forbidden. Hence the function $A - iB$ is not zero. Let $\alpha = \arg(A-iB)$. Then the above equation can be written

$$\cos\alpha\cos 2\tau - \sin\alpha\sin 2\tau = 0 , \text{ or } \cos(\alpha + 2\tau) = 0 .$$

Thus $\tau = -\frac{\alpha}{2}$ + const. and hence $\delta_c(\tau) = -\frac{1}{2}\delta_c(\alpha)$. Therefore

$$j = -\frac{1}{2}\cdot\frac{1}{2\pi}\,\delta_c[\arg(A-iB)] .$$

3. The Main Theorem

3.1 Introduction. We will give two proofs of the main theorem. Our first proof is the proof given by Cohn-Vossen in 1927 and later simplified by Shitomirsky. This proof depends on the surface being analytic and shows that if (λ,μ,ν) define a quadratic form on the sphere satisfying 1.3, 1.4, and 1.6, then it is identically zero. Analyticity can be avoided by more recent achievements in the theory of differential equations.

The second proof is a proof given by Herglotz in 1943 which works for C^3 surfaces. Here we will utilize the Remark of 1.3 that it is sufficient to show that $\lambda\nu - \mu^2 = 0$ and will prove that

$$\iint_{S_0} (\lambda\nu-\mu^2)\,P\,dA = 0$$

where P is a strictly positive function. It follows from this that, since $\lambda\nu - \mu^2 \leq 0$, in fact $\lambda\nu - \mu^2 = 0$.

3.2 Lemma. Let $\lambda\,du^2 + 2\mu\,du dv + \nu\,dv^2$ be the form defined in 1.1 and suppose it satisfies the conclusions of Theorems 1.3, 1.4, and 1.6; i.e.

1) $\lambda\nu-\mu^2 \leqslant 0$ and if $\lambda\nu-\mu^2 = 0$, then $\lambda = \mu = \nu = 0$.

2) There is a positive definite quadratic form (L,M,N) such that

$$-\lambda\nu + \mu^2 = N\lambda - 2M\mu + L\nu$$

3)
$$\lambda_v - \mu_u = 0 \mod J(\lambda,\mu,\nu)$$
$$\mu_v - \nu_u = 0 \mod J(\lambda,\mu,\nu) \quad .$$

Let p be a singularity of the curves

$$\lambda\, du^2 + 2\mu\, du dv + \nu\, dv^2 = 0 \; .$$

If λ, μ , and ν are analytic and not identically zero, then p is an isolated singularity and the index, j , of p is negative.

<u>Proof</u>: By applying a coordinate transformation, we may assume that at p ,

$$(L,M,N) = (1,0,1) \; .$$

Let the Taylor's developments of λ , μ and ν around p be

$$\lambda = \lambda^{(n)} + \dots$$

$$\mu = \mu^{(n)} + \dots$$

$$\nu = \nu^{(n)} + \dots$$

where at least one of $\lambda^{(n)}$, $\mu^{(n)}$, and $\nu^{(n)}$ is not zero. $n > 0$ since p is a singularity. By Condition 2)

$$-\lambda\nu + \mu^2 = N\lambda - 2M\mu + L\nu \; .$$

The left hand side has degree at least $2n$ while on the right hand side, $M\mu$ starts with terms of degree greater than n since $M = 0$ at p while N and L start with 1 . Hence

1)
$$\lambda^{(n)} + \nu^{(n)} = 0 \; .$$

Now, by Condition 3)

$$\lambda_v - \mu_u = a_1\lambda + a_2\mu + a_3\nu$$

$$\mu_v - \nu_u = b_1\lambda + b_2\mu + b_3\nu \; .$$

Since the right hand sides of these equations have no terms of degree $n-1$, it follows that

2)
$$\lambda_v^{(n)} - \mu_u^{(n)} = 0$$

3)
$$\mu_v^{(n)} - \nu_u^{(n)} = 0 \; .$$

But from 1) and 3) we get

3') $\lambda_u^{(n)} + \mu_v^{(n)} = 0$.

Equations 2) and 3') are the Cauchy-Riemann equations for $\lambda^{(n)} - i\mu^{(n)}$. Hence if $w = u + iv$, then

$$\lambda^{(n)} - i\mu^{(n)} = cw^n .$$

Let $\varphi = \lambda - i\mu$. Then

$$\varphi = cw^n + \dots .$$

Then by the argument of VI, 3.4

$$\delta_c(\arg \varphi) = \delta_c(\arg cw^n) = n2\pi .$$

Therefore, by Theorem 2.3

$$j = -\frac{1}{2}\cdot\frac{1}{2\pi}\,\delta_c[\arg(\lambda - i\mu)] = -\frac{n}{2} < 0 .$$

3.3 Theorem. Two isometric ovaloids are congruent.

First Proof: Let $\lambda\,du^2 + 2\mu\,du dv + \nu\,dv^2$ be the form defined in 1.1. Then it can be considered as a form on a sphere and it satisfies properties 1), 2), and 3) of Lemma 3.2. Hence by Lemma 3.2 and Poincaré's Theorem, it follows exactly as in VI, 3.3, 3.4, 2.3 and 2.4 that $\lambda = \mu = \nu \equiv 0$.

Question. Does there exist a tensor $(\lambda,\mu,\nu) \not\equiv (0,0,0)$ on the sphere satisfying properties 1) and 3) of Lemma 3.2 but not property 2) ?

3.4 Definition. Let $q \in E^3$ and let $p(x)$ be the distance from q to the tangent plane to the surface S at the point x . Suppose the origin of the coordinate system of E^3 is at q . Then if X is the position vector of S and N is the inner normal, $p = |XN|$. If S is an ovaloid and q is in the interior of S , then we may write

$$p = -XN$$

and p is strictly positive. p is called the support function of S with respect to q .

3.5 Second Proof: We may assume that the intersection of the interiors of S and S^* is non-empty. Let q be a point in the intersection and let p and p^* be the support functions of S and S^* respectively. Now $EG - F^2 > 0$ and $\frac{\lambda\nu-\mu^2}{EG-F^2}$ is a scalar function. We will prove that

1) $$\iint_{S_0} \frac{\lambda\nu-\mu^2}{EG-F^2}(p+p^*)\,dA = 0 .$$

Since $p + p^* > 0$ and since we know that either $\lambda\nu - \mu^2 < 0$ or $\lambda = \mu = \nu = 0$, this will prove the theorem.

But we will show in fact that 1) holds for every pair of isometric closed surfaces, even if they are not ovaloids. For

$$\frac{\lambda\nu-\mu^2}{EG-F^2} = \frac{1}{EG-F^2}\begin{vmatrix} L^* - L & M^* - M \\ M^* - M & N^* - N \end{vmatrix}$$

$$= 2K - 2K'$$

where K is the common Gauss curvature of S and S^* and

$$2K' = \frac{LN^*-2MM^*+NL^*}{EG-F^2}\ .$$

Hence 1) is equivalent to

1') $$\iint_{S_0} (K-K')(p+p^*)\,dA = 0\ .$$

Now the integrand can be rewritten

$$(K-K')(p+p^*) = (Kp-H) + (Kp^*-H^*) - (K'p-H^*) - (K'p^*-H)\ .$$

It is sufficient to prove that

2) $$\iint (K'p-H^*)\,dA = 0$$

since then the other integrals are also zero either by symmetry or by indentifying the surfaces S and S^*. This formula is quite analogous to the well-known formula of Minkowski for ovaloids, that

3) $$\iint_{S_0} (Kp-H)\,dA = 0\ .$$

Formula 3) follows with the aid of Stokes Theorem, since it can be shown that if R is a region of a surface with boundary B, then

4) $$-2\iint_R (Kp-H)\,dA = \oint_B (X,N,dN)\ .$$

Formula 3) follows immediately from this since $\oint(X,N,dN)$ does not depend on the coordinate system and hence cancels out when the integration is extended over a closed surface.

In 4), $N_i = -\ell_i^j X_j$ (see I, 8.2) and hence $dN = -\ell_i^j X_j du^i$. Formula 2) follows from the analogous expression

5) $$-2\iint_R (K'p-H^*)\,dA = \oint_B (X,N,\Gamma)$$

where $\Gamma = -\ell^{*j}_i X_j du^i$. It is easy to see that again $\oint_B (X,N,\Gamma)$ is independent of the coordinate system and therefore

$$\iint_{S_0} (K'p-H^*)\,dA = 0 .$$

<u>Exercise</u>: Derive formulas 4) and 5) above using the techniques of exterior differentiation; i.e., show

a) $$d(X,N,dN) = -2(Kp-H)\,dA$$

b) $$d(X,N,\Gamma) = -2(K'p-H^*)\,dA .$$

The first is easy since $ddN = 0$. In the second, it is not true that $d\Gamma = 0$ but we do have that $d\Gamma\cdot X_k = 0$, $k = 1,2$, which helps to give the formula. In the expression for $d\Gamma\cdot X_k$ one can replace X by X^*, using the isometry between the two surfaces. Therefore one has $d\Gamma\cdot X_k = (ddN^*)\cdot X_k^* = 0$.

<u>3.6 Generalizations</u>. The theorem is certainly not true in general if one removes the restriction that the surfaces be ovaloids. The illustration below gives two C^∞ surfaces of revolution which are obvious isometric but not congruent.

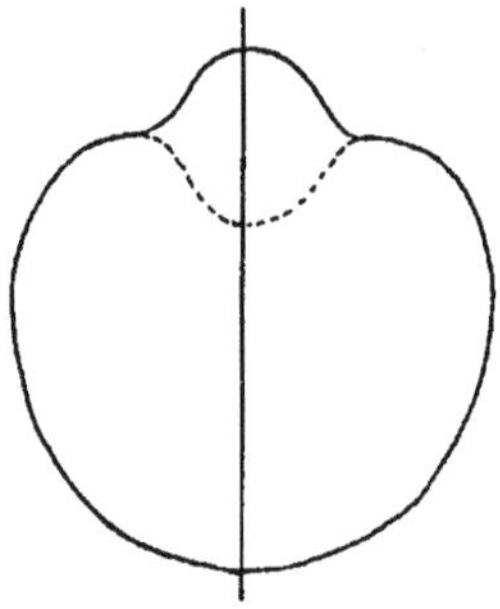

E. Rembs (Math. Zeitschrift, Vol. 56 (1952) p. 274) has given examples of analytic surfaces which are isometric but such that the isometry between them is not a congruence.

A. Alexandrov has proved (in Russian) however, that if S and S^* are simple closed, analytic surfaces (of arbitrary genus) such that

$$\iint_{K>0} K\,dA = 4\pi$$

then the congruence theorem holds. The tori of revolution give an example of this situation.

Our theorem can be restated as follows: Given an abstract closed surface S_0 with a Riemannian metric g_{ij} such that $K>0$, then there exists at most one realization of S_0 with this metric in E^3

(modulo Euclidean motions, of course). It can be shown in fact that there exists exactly one such realization in E^3 . A proof was sketched but not completed by H. Weyl in 1916. One later proof has been given by Alexandrov and Pogorelov, and another proof by Nirenberg. These proofs also contain uniqueness proofs and hence give alternative proofs of our theorem.

CHAPTER IX

Singularities of Surfaces with Constant Negative Gauss Curvature

1. Singularities

1.1 Introduction. In this chapter we shall be concerned with (open) surfaces and their imbeddings in E^3. The definition of an open surface is identical with definition II, 1.1 except that condition 1) that S be compact is no longer true. We will show that a surface with constant negative Gauss curvature cannot be imbedded as a general (open) surface in E^3 without singularities (in a sense to be defined below). The first proof of this was given by Hilbert (~ 1900) for analytic surfaces. Our proof works for C^3 surfaces and the theorem is still true for C^2 surfaces. However, Kuiper has given a C^1 isometric imbedding of the hyperbolic plane in E^3 without singularities. For details see N.H. Kuiper, on C^1-isometric Imbeddings I and II; Indagationes Mathematicae, Vol. 17 (1955) pp 545-556 and pp 683-689.

1.2 The hyperbolic plane. Consider the upper half plane of the u-v-plane (i.e., $v > 0$) with the metric $ds^2 = \dfrac{du^2 + dv^2}{v^2}$. This surface

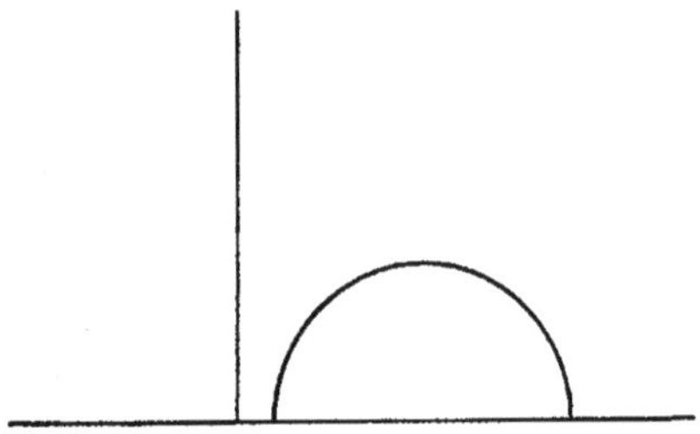

is called the hyperbolic plane. It is easy to show that geodesics on this surface are either straight lines perpendicular to the u-axis or semi-circles with their center on the u-axis.

The first fundamental form can be rewritten

$$ds^2 = \left(\frac{dv}{v}\right)^2 + \frac{1}{v^2}\, du^2 .$$

Let $\bar{v} = \log v$. Then

$$ds^2 = d\bar{v}^2 + e^{-2\bar{v}} du^2 = d\bar{v}^2 + g^2 du^2$$

where $g = e^{-\bar{v}}$. Hence by I, 6.3,

$$K = -\frac{g_{\bar{v}\bar{v}}}{g} \equiv -1 .$$

Hence the hyperbolic plane is a surface with Gauss curvature equal to -1 .

1.3 Examples in E^3. Surfaces of revolution with constant negative Gauss curvature and with singularities are illustrated below:

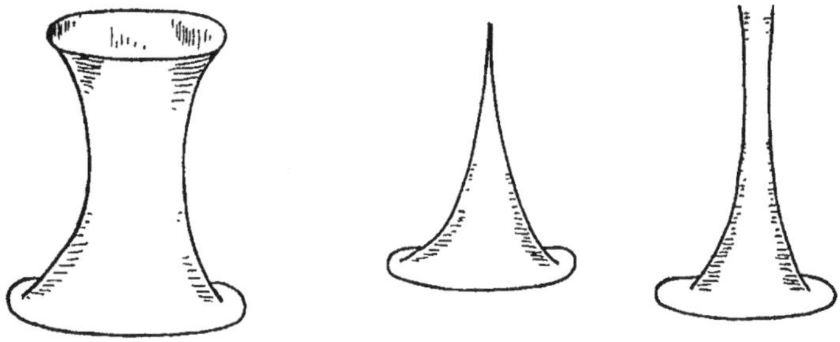

Here one of the principle curvatures is infinite on the singular lines, but the normals are continuous.

1.4 Discussion. One of the main problems is to give a satisfactory definition of singularity. We wish to discuss singularities which arise as a property of the imbedding of an abstract surface S_o in E^3. Therefore, we are not concerned with "singularities" which may occur on an abstract surface (e.g., singularities of a metric, etc.). Consequently, we shall assume that all points of an abstract surface are regular points. As a first attempt at a definition we may say that a point $p \in E^3$ is a singularity of S if $p \notin S$ but if $p \in \bar{S}$ (the closure of S). Now it is not sufficient to require only that there is a sequence $(p_n: p_n \in S)$ which converges to p_o. For, consider a thin strip A, which spirals an infinite number of times around a point p_o, as in a). Then there is clearly a sequence $p_n \in A$ which converges to p_o, but we do not wish to consider p_o as a singularity of A. Thus we should like to require that there actually be a curve of finite length converging to a singularity. But as b) illustrates this is also

a)

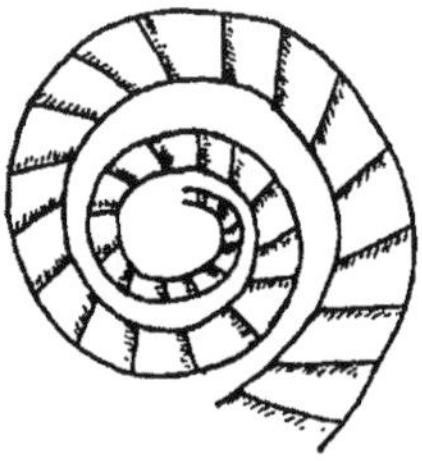

b)

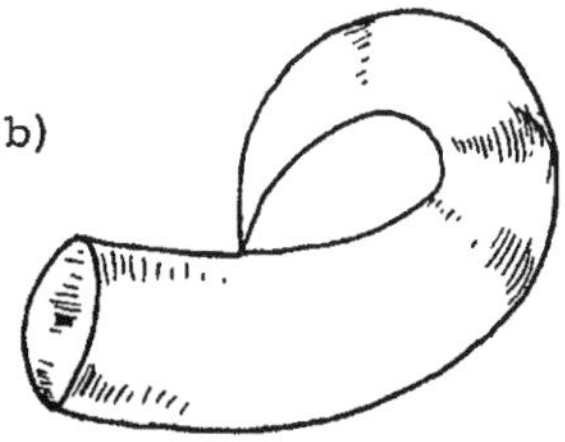

not a satisfactory definition, for here the singularity lies in a point which is also a regular point of the surface. The following definition does turn out to be satisfactory.

1.5 Definition. A surface S_o is called complete (with respect to a Riemannian geometry) if every half-open curve of finite length on S_o (parametized by $0 \leq t < 1$) has an end point (at $t = 1$) on S_o.

If S_o is complete, then we say that an imbedding of S_o in E^3 has no singularities.

A singularity is defined as follows:

Suppose $C_o(t)$, $0 \leq t < 1$ is a divergent curve of finite length on S_o (We will usually consider the metric on S_o induced by the imbedding in E^3); i.e., if $t_n \to 1$, then $C_o(t_n)$ has no limit point on S_o. Let $C(t)$ be the image of C_o in E^3 under an isometric imbedding of S_o in E^3. Then C is a curve of finite length in E^3 and hence converges to a point $p \in E^3$. Then we say that a singularity of the image of S_o in E^3 lies in the point p. (As in 1.4, there may also be regular points of S_o at p.)

1.6 Further discussion. One immediate consequence of our definition is that if every divergent curve on S_o has infinite length and if S_o has an isometric imbedding in E^3, then the imbedded surface has no singularities. It is easy to see that the hyperbolic plane satisfies the property that every divergent curve has infinite length. Hence if the hyperbolic plane had an isometric imbedding in E^3, the imbedding would have no singularities. However, our main theorem in this chapter will tell us that any isometric imbedding of a surface with constant negative curvature necessarily has singularities. Therefore, there is no isometric imbedding of the hyperbolic plane in E^3.

Our definition allows certain more or less trivial singularities which are really not relevant to the discussion. For suppose S_o' is a proper, open subset of S_o. Then S_o' has boundary points relative to S_o. It is clear that under an isometric imbedding of S_o' these boundary points are singularities, while under an imbedding of S_o, they are not. Such singularities which can be removed by an extension of the original surface are called ordinary singular points (or removable singularities). We will always assume that such extensions have been carried out since we are concerned with "intrinsic" singularities which cannot be eliminated. It can be shown by Zorn's lemma that any surface can be extended in this sense to a surface which cannot be extended any farther.

2. Tschebyscheff Nets

2.1 Definition. Let R be a piece of a surface on which there is defined a net consisting of two distinct families of regular curves. We will introduce local parameters u and v along these lines so that we may speak of u-lines and v-lines. Choose a positive orientation of R and let ω be the positive angle through which a u-line must be rotated to become tangent to a v-line. We will always assume that the u and v lines are choosen to satisfy $0 < \omega < \pi$.

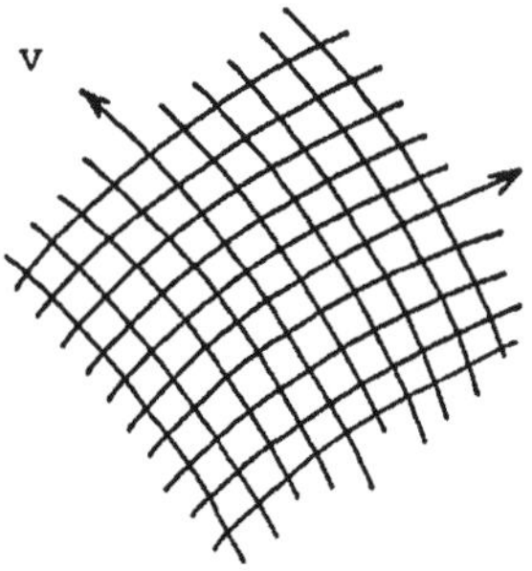

If each rectangle whose sides are u-lines and v-lines has the property, that opposite sides have the same length, then the net is called a Tschebyscheff net.

On a Tschebyscheff net we can introduce parameters u and v such that u and v measure arc length respectively on the u-lines and the v-lines. In such a coordinate system the first fundamental form becomes

$$ds^2 = du^2 + 2F\,dudv + dv^2$$

where $F = \cos\omega$. Conversely, it is clear that the lines u = constant and v = constant in such a coordinate system form a Tschebyscheff net. We will call this coordinate system a Tschebyscheff coordinate system.

2.2 Lemma. In a Tschebyscheff coordinate system the Gauss curvature K is given by

$$\omega_{uv} = -K \sin\omega$$

Proof: For such a coordinate system we have

$$EG-F^2 = \sin^2\omega .$$

Let $W = \sqrt{EG-F^2}$. Then by the Theorema Egregium

$$K = \frac{1}{2W}\left[\left(\frac{F_u}{W}\right)_v + \left(\frac{F_v}{W}\right)_u\right] = -\frac{\omega_{uv}}{\sin\omega}$$

Thus $\omega_{uv} = -K \sin \omega$.

2.3 Lemma. If R_0 is a rectangle as in 2.1, then

$$\left| \iint_{R_0} K \, dA \right| < 2\pi$$

Proof: Let R_0 be a rectangle whose sides are u-lines and v-lines. Then since

$$dA = \sin\omega \, du dv$$

we have $K\, dA = -\omega_{uv} du dv$.

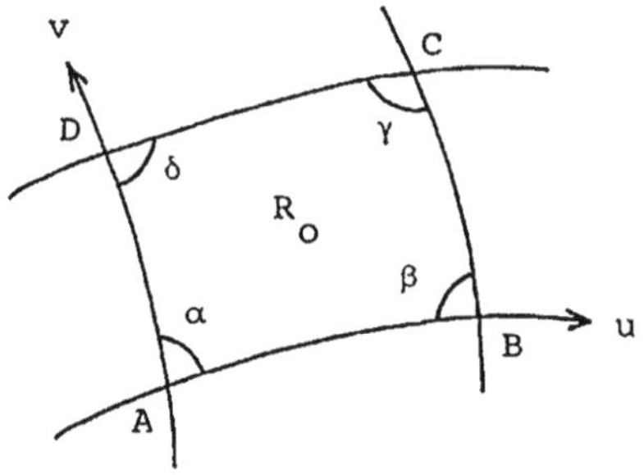

Hence
$$\iint_{R_0} K \, dA = - \iint \omega_{uv} du dv = - \oint \omega_v dv$$

by Stokes Theorem, where the line integral is taken in the positive direction over the boundary of R_0 . Thus, using the notation of the illustration,

$$\iint_{R_0} K \, dA = - \int_B^C \omega_v dv + \int_A^D \omega_v dv$$

$$= [-\gamma + (\pi - \beta)] + [(\pi - \delta) - \alpha]$$

$$= 2\pi - (\alpha + \beta + \gamma + \delta) \; .$$

Because of $0 < \alpha, \beta, \gamma, \delta < \pi$, the value of the integral lies between -2π and 2π .

Remark: The above lemma used only the concepts of Riemannian geometry and did not depend on properties of an imbedding in a Euclidean space. This is not true for the following theorem, however.

2.4 Theorem. Let R be a region of a surface in E^3 on which $K \equiv -1$. Then the asymptotic lines on R form a Tschebyscheff net.

Proof: If $K < 0$ then the asymptotic lines are the solutions of

1) $L\, du^2 + 2M\, du dv + N\, dv^2 = 0$. If we make the asymptotic lines the u and v lines then we must have that $du = 0$ and $dv = 0$ are solutions of equation 1); i.e. $L = N = 0$. In such a coordinate system the

Codazzi equations reduce to

2) $$M_u = AM$$
$$M_v = BM$$

where

$$A = \frac{[\frac{1}{2}(EG-F^2)_u + FE_v - EG_u]}{EG-F^2}$$
$$B = \frac{[\frac{1}{2}(EG-F^2)_v + FG_u - GE_v]}{EG-F^2}$$

(See, for example, Blaschke (3rd ed.) p. 117, Formula 139). Equations 2) can be rewritten

2') $$(M^2)_u = 2\ AM^2$$
$$(M^2)_v = 2\ BM^2 .$$

But if $K = -1$, then
$$\frac{-M^2}{EG-F^2} = -1 .$$

or $M^2 = EG-F^2$. Hence, substituting in 2'),

$$(EG-F^2)_u = (EG-F^2)_u + 2(FE_v - EG_u) ,$$

or $$EG_u - FE_v = 0$$

Similarly $$-FG_u + GE_v = 0 .$$

But since $EG-F^2 \neq 0$, these equations is satisfied if and only if

3) $$E_v = 0 \quad \text{and} \quad G_u = 0 .$$

Therefore we have

$$E = E(u) \quad \text{and} \quad G = G(v) .$$

Let $\bar{u}$ and $\bar{v}$ be coordinates such that

$$d\bar{u} = \sqrt{E(u)}\,du$$
$$d\bar{v} = \sqrt{G(v)}\,dv .$$

Then the first fundamental form becomes

$$ds^2 = d\bar{u}^2 + 2F\,d\bar{u}d\bar{v} + d\bar{v}^2 .$$

Hence the $\bar{u}$ and $\bar{v}$-lines form a Tschebyscheff net.

3. The Main Theorem

3.1 In this section we will show that any surface S in E^3 with constant negative Gauss curvature has singularities. To do this we must show that if S_o is a parameter surface for S, then S_o is not complete. We will show that if the canonical Tschebyscheff net on S is considered as a net on S_o, then at least one curve of the net contains a divergent arc of finite length. From this it follows that S has at least one singular point. In Section 4 we will discuss (without proofs) the kinds of singularities which actually appear.

3.2 Definition: For the canonical Tschebyscheff net on a region R of a surface with $K = -1$, we have, using the notations and conventions of the preceding section, that

$$\omega_{uv} = -K \sin \omega = \sin \omega > 0 .$$

This equation clearly does not depend on the orientations of the u and v-lines.

Let ℓ be a v-line of the net and let $p \in \ell$. Now ω_u is a monotone function on ℓ since $(\omega_u)_v > 0$, and hence there is at most one point on ℓ where $\omega_u = 0$. Therefore, we may assume that $\omega_u(p) \neq 0$. Let the positive u-direction be the direction such that

$$\omega_u(p) > 0 .$$

This determines the positive v-direction if we require that a positive rotation of angle ω (where $0 < \omega < \pi$) carry the positive u-direction into the positive v-direction. This direction on the v-lines is called the distinguished direction.

It is easy to see that this direction is independent of the orientation of R; for, suppose the opposite orientation of R had been chosen. Let $\bar{\omega}$ be the corresponding angle. Then $\bar{\omega} = \pi - \omega$, so

$$\bar{\omega}_u = - \omega_u$$

and hence in the above discussion we must choose the opposite u-direction. But then a positive rotation (with this orientation of R) obviously carries this u-direction into exactly the same v-direction as above.

3.3 Theorem. A surface S in E^3 with constant negative Gauss curvature has singular points.

Proof: Let S_o be a parameter surface for S and consider the canonical Tschebyscheff net on S as a net on S_0 . Let $p \in S_0$ be a point where $\omega_u \neq 0$ and let the positive u and v directions be chosen as in 3.2. Let q be a point on the positive u-line through p . We will be concerned with the region R above (in the positive, distinguished v-direction from) the u-line through p and q . Since we have a Tschebyscheff net on this region, at least the lower part of this region corresponds to a rectangle R_0 in the u-v-plane. It is clear that for a sufficiently small positive number V , it is possible to measure

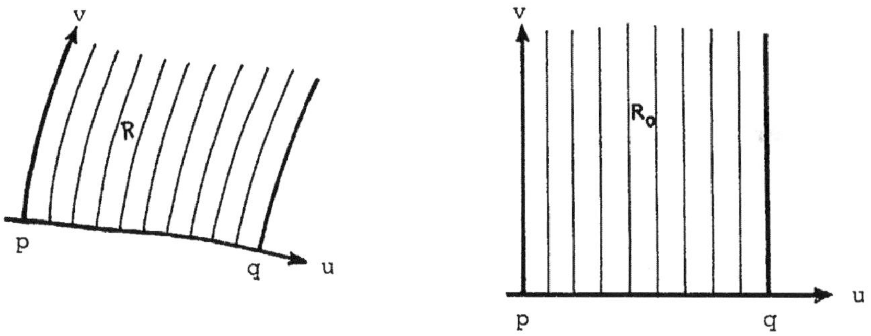

a distance V along a v-line in the distinguished v-direction starting from the u-line through p and q . This can be done uniformly for each v-line in R . To prove our theorem it is sufficient to show that there is a finite least upper bound V^* to the distances which can be measured along all v-lines in R . For suppose V^* is such a bound. Then there must be a v-line in R along which it is not possible to measure the distance V^* . Hence the arc of this v-line starting from the curve between p and q is a half-open divergent curve of finite length and therefore S has a singular point.

To prove that $V^* < \infty$ we proceed as follows: let U be the length of the u-line between p and q and let U' satisfy

$$0 < U' < U .$$

Choose p' and q' so that

$$p < p' < q' < q$$

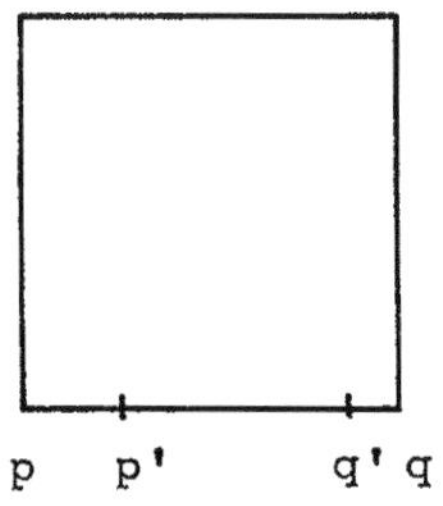

and so that U' is the length of $\overline{p'q'}$. Since $\omega_u(p) > 0$ we may assume that q is close enough to p so that $\omega_u > 0$ on all of $\overline{pq}$. Then ω is an increasing function on $\overline{pq}$ and

$$\omega(p') - \omega(p) > 0$$

$$\omega(q) - \omega(q') > 0 \quad .$$

Hence we can find an $\varepsilon > 0$ which is smaller than both of these quantities. (Since $0 < \omega < \pi$, it follows that $\varepsilon < \frac{\pi}{2}$).

Now, consider a rectangle of height V over $\overline{pq}$. Along the edge above q, $\omega < \pi$. Further, if q_1 and q_1' are situated as illustrated then, since $(\omega_u)_v > 0$, we have

$$\omega(q_1) - \omega(q_1') = \int_{q_1'}^{q_1} \omega_u du$$

$$> \int_{q'}^{q} \omega_u du > \varepsilon \; .$$

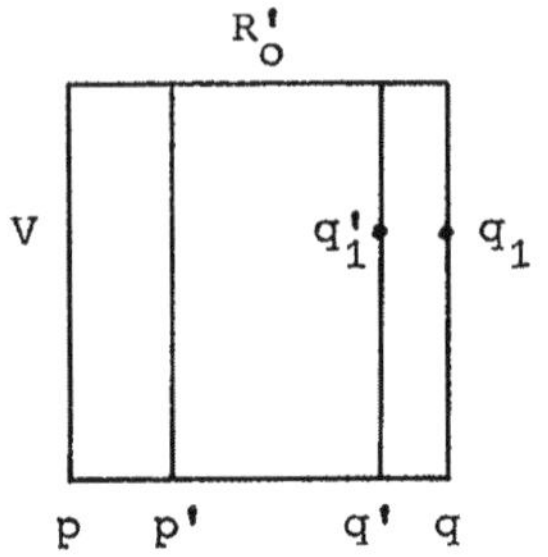

A similar inequality holds for points above p and p'. Therefore, if r is any point in the smaller rectangle R_o' (above $\overline{p'q'}$), it follows that

$$\varepsilon < \omega(r) < \pi - \varepsilon$$

and hence

$$\sin \omega > \sin \varepsilon \quad .$$

Consequently, by Lemma 2.3, and since

$$-K\,dA = \sin \omega \, dudv$$

we have

$$2\pi > \iint_{R_o} \sin \omega \, dudv > \sin \varepsilon \iint_{R_o'} dudv = U'V \sin \varepsilon \; .$$

Hence

$$V < \frac{2\pi}{U' \sin \varepsilon}$$

so $V^* = \dfrac{2\pi}{U' \sin \varepsilon}$ is an upper bound, which proves the theorem.

4. Further Details and Generalizations

4.1 Singularities. M.H. Amsler has shown that on an analytic surface with constant negative Gauss curvature, there exists a C^{∞}-curve consisting entirely of singular points. However, his proof does not hold for C^n surfaces. For details see

Amsler, M.H., Des surfaces à courbure négative constante dans l'espace à trois dimensions et de leurs singularités; Mathematische Annalen, 130(1955) pp. 234-256.

4.2 Constant Positive Curvature. It can be shown that if S_o is an abstract surface on which there is defined a Riemann metric such that S_o is complete and $K \equiv 1$, then S_o is compact. It follows immediately from this that the only surface with constant positive Gauss curvature in E^3 without singularities is a sphere.

4.3 Strictly Positive or Strictly Negative Curvature. E. Heinz has proved the following theorems concerning surfaces in E^3 given by a function $z = z(x,y)$ which is defined and of class C^2 in the circle $x^2 + y^2 < R^2$.

1) If $|H| \geqslant \alpha > 0$, then $R \leqslant \frac{1}{\alpha}$

2) If $K \geqslant \alpha > 0$, then $R \leqslant \sqrt{\frac{1}{\alpha}}$

3) If $K \leqslant -\alpha < 0$, then $R \leqslant e\sqrt{\frac{3}{\alpha}}$

For details see:

E. Heinz, Ueber Flächen mit eineindeutiger Projektion auf eine Ebene, deren Krümmungen durch Ungleichungen eingeschränkt sind; Mathematische Annalen, 129(1955) p. 451-454.

It can also be shown that the theorem stated in 4.2 is true if one only requires that $K \geqslant K_o > 0$. It follows from this that if S is such a surface in E^3 without singularities, then S is an ovaloid.

If we only require that $K > 0$, then not too much can be said. However, Stoker has shown that a part of Hadamard's Theorem (IV, 1.4) is still true; namely:

If S is a complete surface in E^3 with $K > 0$, then S has no self-intersections and S is the boundary of a convex set.

It follows from this that the spherical map is 1-1 and $\int_S K\,dA \leqslant 2\pi$.

<u>4.4 The Curvature Integral</u> (See III, 2.3). In general, for an open surface S_o, $\iint_{S_o} K\,dA$ does not exist. However, if S_o is "complete" it can be shown that there exists an increasing sequence of regions G_n, each of finite area, such that $S_o = \cup G_n$ and such that

$$\varlimsup_{n\to\infty} \iint_{G_n} K\,dA \leq 2\pi(1-p_1)$$

where p_1 is the first Betti number of S_o. For a closed surface, we had

$$\iint K\,dA = 2\pi(2-p_1)$$

But for any surface, $p_0 = 1$ and $p_2 = 1$ if the surface is closed and $p_2 = 0$ if the surface is open. Hence it is always true that for each complete surface (closed or open)

$$\varlimsup_{n\to\infty} \iint_{G_n} K\,dA \leq 2\pi(p_0 - p_1 + p_2) \,.$$

For details, cf. S. Cohn-Vossen, Compositio Mathematica 2 (1935).

LECTURE NOTES IN MATHEMATICS

Edited by A. Dold and B. Eckmann

Some general remarks on the publication of monographs and seminars

In what follows all references to monographs, are applicable also to multiauthorship volumes such as seminar notes.

§1. Lecture Notes aim to report new developments - quickly, informally, and at a high level. Monograph manuscripts should be reasonably self-contained and rounded off. Thus they may, and often will, present not only results of the author but also related work by other people. Furthermore, the manuscripts should provide sufficient motivation, examples and applications. This clearly distinguishes Lecture Notes manuscripts from journal articles which normally are very concise. Articles intended for a journal but too long to be accepted by most journals, usually do not have this "lecture notes" character. For similar reasons it is unusual for Ph.D. theses to be accepted for the Lecture Notes series.

Experience has shown that English language manuscripts achieve a much wider distribution.

§2. Manuscripts or plans for Lecture Notes volumes should be submitted either to one of the series editors or to Springer-Verlag, Heidelberg. These proposals are then refereed. A final decision concerning publication can only be made on the basis of the complete manuscripts, but a preliminary decision can usually be based on partial information: a fairly detailed outline describing the planned contents of each chapter, and an indication of the estimated length, a bibliography, and one or two sample chapters - or a first draft of the manuscript. The editors will try to make the preliminary decision as definite as they can on the basis of the available information.

§3. Lecture Notes are printed by photo-offset from typed copy delivered in camera-ready form by the authors. Springer-Verlag provides technical instructions for the preparation of manuscripts, and will also, on request, supply special staionery on which the prescribed typing area is outlined. Careful preparation of the manuscripts will help keep production time short and ensure satisfactory appearance of the finished book. Running titles are not required; if however they are considered necessary, they should be uniform in appearance. We generally advise authors not to start having their final manuscripts specially tpyed beforehand. For professionally typed manuscripts, prepared on the special stationery according to our instructions, Springer-Verlag will, if necessary, contribute towards the typing costs at a fixed rate.

The actual production of a Lecture Notes volume takes 6-8 weeks.

.../...

§4. Final manuscripts should contain at least 100 pages of mathema tical text and should include

- a table of contents
- an informative introduction, perhaps with some historical re marks. It should be accessible to a reader not particularly familiar with the topic treated.
- a subject index; this is almost always genuinely helpful for the reader.

§5. Authors receive a total of 50 free copies of their volume, bu no royalties. They are entitled to purchase further copies o their book for their personal use at a discount of 33.3 % other Springer mathematics books at a discount of 20 % directl from Springer-Verlag.

Commitment to publish is made by letter of intent rather than b signing a formal contract. Springer-Verlag secures the copyrigh for each volume.

l. 1232: P.C. Schuur, Asymptotic Analysis of Soliton Problems. VIII, 0 pages. 1986.

l. 1233: Stability Problems for Stochastic Models. Proceedings, 85. Edited by V.V. Kalashnikov, B. Penkov and V.M. Zolotarev. VI, 3 pages. 1986.

l. 1234: Combinatoire énumérative. Proceedings, 1985. Edité par Labelle et P. Leroux. XIV, 387 pages. 1986.

l. 1235: Séminaire de Théorie du Potentiel, Paris, No. 8. Directeurs: . Brelot, G. Choquet et J. Deny. Rédacteurs: F. Hirsch et G. okobodzki. III, 209 pages. 1987.

l. 1236: Stochastic Partial Differential Equations and Applications. oceedings, 1985. Edited by G. Da Prato and L. Tubaro. V, 257 ges. 1987.

l. 1237: Rational Approximation and its Applications in Mathematics d Physics. Proceedings, 1985. Edited by J. Gilewicz, M. Pindor and . Siemaszko. XII, 350 pages. 1987.

l. 1238: M. Holz, K.-P. Podewski and K. Steffens, Injective Choice nctions. VI, 183 pages. 1987.

l. 1239: P. Vojta, Diophantine Approximations and Value Distribu- n Theory. X, 132 pages. 1987.

l. 1240: Number Theory, New York 1984–85. Seminar. Edited by V. Chudnovsky, G.V. Chudnovsky, H. Cohn and M.B. Nathanson. 324 pages. 1987.

l. 1241: L. Gårding, Singularities in Linear Wave Propagation. III, 5 pages. 1987.

l. 1242: Functional Analysis II, with Contributions by J. Hoffmann- rgensen et al. Edited by S. Kurepa, H. Kraljević and D. Butković. VII, 2 pages. 1987.

ol. 1243: Non Commutative Harmonic Analysis and Lie Groups. roceedings, 1985. Edited by J. Carmona, P. Delorme and M. Vergne. , 309 pages. 1987.

ol. 1244: W. Müller, Manifolds with Cusps of Rank One. XI, 158 ages. 1987.

ol. 1245: S. Rallis, L-Functions and the Oscillator Representation. VI, 239 pages. 1987.

ol. 1246: Hodge Theory. Proceedings, 1985. Edited by E. Cattani, F. uillén, A. Kaplan and F. Puerta. VII, 175 pages. 1987.

ol. 1247: Séminaire de Probabilités XXI. Proceedings. Edité par J. zéma, P.A. Meyer et M. Yor. IV, 579 pages. 1987.

ol. 1248: Nonlinear Semigroups, Partial Differential Equations and ttractors. Proceedings, 1985. Edited by T.L. Gill and W.W. Zachary. , 185 pages. 1987.

ol. 1249: I. van den Berg, Nonstandard Asymptotic Analysis. IX, 187 ages. 1987.

ol. 1250: Stochastic Processes – Mathematics and Physics II. roceedings 1985. Edited by S. Albeverio, Ph. Blanchard and L. Streit. I, 359 pages. 1987.

ol. 1251: Differential Geometric Methods in Mathematical Physics. roceedings, 1985. Edited by P.L. García and A. Pérez-Rendón. VII, 00 pages. 1987.

ol. 1252: T. Kaise, Représentations de Weil et GL_2 Algèbres de ivision et GL_n. VII, 203 pages. 1987.

ol. 1253: J. Fischer, An Approach to the Selberg Trace Formula via e Selberg Zeta-Function. III, 184 pages. 1987.

ol. 1254: S. Gelbart, I. Piatetski-Shapiro, S. Rallis. Explicit Construc- ons of Automorphic L-Functions. VI, 152 pages. 1987.

ol. 1255: Differential Geometry and Differential Equations. Proceed- gs, 1985. Edited by C. Gu, M. Berger and R.L. Bryant. XII, 243 ages. 1987.

ol. 1256: Pseudo-Differential Operators. Proceedings, 1986. Edited y H.O. Cordes, B. Gramsch and H. Widom. X, 479 pages. 1987.

ol. 1257: X. Wang, On the C*-Algebras of Foliations in the Plane. V, 65 pages. 1987.

ol. 1258: J. Weidmann, Spectral Theory of Ordinary Differential Operators. VI, 303 pages. 1987.

Vol. 1259: F. Cano Torres, Desingularization Strategies for Three-Dimensional Vector Fields. IX, 189 pages. 1987.

Vol. 1260: N.H. Pavel, Nonlinear Evolution Operators and Semi-groups. VI, 285 pages. 1987.

Vol. 1261: H. Abels, Finite Presentability of S-Arithmetic Groups. Compact Presentability of Solvable Groups. VI, 178 pages. 1987.

Vol. 1262: E. Hlawka (Hrsg.), Zahlentheoretische Analysis II. Seminar, 1984–86. V, 158 Seiten. 1987.

Vol. 1263: V.L. Hansen (Ed.), Differential Geometry. Proceedings, 1985. XI, 288 pages. 1987.

Vol. 1264: Wu Wen-tsün, Rational Homotopy Type. VIII, 219 pages. 1987.

Vol. 1265: W. Van Assche, Asymptotics for Orthogonal Polynomials. VI, 201 pages. 1987.

Vol. 1266: F. Ghione, C. Peskine, E. Sernesi (Eds.), Space Curves. Proceedings, 1985. VI, 272 pages. 1987.

Vol. 1267: J. Lindenstrauss, V.D. Milman (Eds.), Geometrical Aspects of Functional Analysis. Seminar. VII, 212 pages. 1987.

Vol. 1268: S.G. Krantz (Ed.), Complex Analysis. Seminar, 1986. VII, 195 pages. 1987.

Vol. 1269: M. Shiota, Nash Manifolds. VI, 223 pages. 1987.

Vol. 1270: C. Carasso, P.-A. Raviart, D. Serre (Eds.), Nonlinear Hyperbolic Problems. Proceedings, 1986. XV, 341 pages. 1987.

Vol. 1271: A.M. Cohen, W.H. Hesselink, W.L.J. van der Kallen, J.R. Strooker (Eds.), Algebraic Groups Utrecht 1986. Proceedings. XII, 284 pages. 1987.

Vol. 1272: M.S. Livšic, L.L. Waksman, Commuting Nonselfadjoint Operators in Hilbert Space. III, 115 pages. 1987.

Vol. 1273: G.-M. Greuel, G. Trautmann (Eds.), Singularities, Repre-sentation of Algebras, and Vector Bundles. Proceedings, 1985. XIV, 383 pages. 1987.

Vol. 1274: N. C. Phillips, Equivariant K-Theory and Freeness of Group Actions on C*-Algebras. VIII, 371 pages. 1987.

Vol. 1275: C.A. Berenstein (Ed.), Complex Analysis I. Proceedings, 1985–86. XV, 331 pages. 1987.

Vol. 1276: C.A. Berenstein (Ed.), Complex Analysis II. Proceedings, 1985–86. IX, 320 pages. 1987.

Vol. 1277: C.A. Berenstein (Ed.), Complex Analysis III. Proceedings, 1985–86. X, 350 pages. 1987.

Vol. 1278: S.S. Koh (Ed.), Invariant Theory. Proceedings, 1985. V, 102 pages. 1987.

Vol. 1279: D. Ieşan, Saint-Venant's Problem. VIII, 162 Seiten. 1987.

Vol. 1280: E. Neher, Jordan Triple Systems by the Grid Approach. XII, 193 pages. 1987.

Vol. 1281: O.H. Kegel, F. Menegazzo, G. Zacher (Eds.), Group Theory. Proceedings, 1986. VII, 179 pages. 1987.

Vol. 1282: D.E. Handelman, Positive Polynomials, Convex Integral Polytopes, and a Random Walk Problem. XI, 136 pages. 1987.

Vol. 1283: S. Mardešić, J. Segal (Eds.), Geometric Topology and Shape Theory. Proceedings, 1986. V, 261 pages. 1987.

Vol. 1284: B.H. Matzat, Konstruktive Galoistheorie. X, 286 pages. 1987.

Vol. 1285: I.W. Knowles, Y. Saitō (Eds.), Differential Equations and Mathematical Physics. Proceedings, 1986. XVI, 499 pages. 1987.

Vol. 1286: H.R. Miller, D.C. Ravenel (Eds.), Algebraic Topology. Proceedings, 1986. VII, 341 pages. 1987.

Vol. 1287: E.B. Saff (Ed.), Approximation Theory, Tampa. Proceed-ings, 1985–1986. V, 228 pages. 1987.

Vol. 1288: Yu. L. Rodin, Generalized Analytic Functions on Riemann Surfaces. V, 128 pages, 1987.

Vol. 1289: Yu. I. Manin (Ed.), K-Theory, Arithmetic and Geometry. Seminar, 1984–1986. V, 399 pages. 1987.

Vol. 1290: G. Wüstholz (Ed.), Diophantine Approximation and Transcendence Theory. Seminar, 1985. V, 243 pages. 1987.

Vol. 1291: C. Mœglin, M.-F. Vignéras, J.-L. Waldspurger, Correspondances de Howe sur un Corps p-adique. VII, 163 pages. 1987

Vol. 1292: J.T. Baldwin (Ed.), Classification Theory. Proceedings, 1985. VI, 500 pages. 1987.

Vol. 1293: W. Ebeling, The Monodromy Groups of Isolated Singularities of Complete Intersections. XIV, 153 pages. 1987.

Vol. 1294: M. Queffélec, Substitution Dynamical Systems – Spectral Analysis. XIII, 240 pages. 1987.

Vol. 1295: P. Lelong, P. Dolbeault, H. Skoda (Réd.), Séminaire d'Analyse P. Lelong – P. Dolbeault – H. Skoda. Seminar, 1985/1986. VII, 283 pages. 1987.

Vol. 1296: M.-P. Malliavin (Ed.), Séminaire d'Algèbre Paul Dubreil et Marie-Paule Malliavin. Proceedings, 1986. IV, 324 pages. 1987.

Vol. 1297: Zhu Y.-l., Guo B.-y. (Eds.), Numerical Methods for Partial Differential Equations. Proceedings. XI, 244 pages. 1987.

Vol. 1298: J. Aguadé, R. Kane (Eds.), Algebraic Topology, Barcelona 1986. Proceedings. X, 255 pages. 1987.

Vol. 1299: S. Watanabe, Yu.V. Prokhorov (Eds.), Probability Theory and Mathematical Statistics. Proceedings, 1986. VIII, 589 pages. 1988.

Vol. 1300: G.B. Seligman, Constructions of Lie Algebras and their Modules. VI, 190 pages. 1988.

Vol. 1301: N. Schappacher, Periods of Hecke Characters. XV, 160 pages. 1988.

Vol. 1302: M. Cwikel, J. Peetre, Y. Sagher, H. Wallin (Eds.), Function Spaces and Applications. Proceedings, 1986. VI, 445 pages. 1988.

Vol. 1303: L. Accardi, W. von Waldenfels (Eds.), Quantum Probability and Applications III. Proceedings, 1987. VI, 373 pages. 1988.

Vol. 1304: F.Q. Gouvêa, Arithmetic of p-adic Modular Forms. VIII, 121 pages. 1988.

Vol. 1305: D.S. Lubinsky, E.B. Saff, Strong Asymptotics for Extremal Polynomials Associated with Weights on $\mathbb{R}$. VII, 153 pages. 1988.

Vol. 1306: S.S. Chern (Ed.), Partial Differential Equations. Proceedings, 1986. VI, 294 pages. 1988.

Vol. 1307: T. Murai, A Real Variable Method for the Cauchy Transform, and Analytic Capacity. VIII, 133 pages. 1988.

Vol. 1308: P. Imkeller, Two-Parameter Martingales and Their Quadratic Variation. IV, 177 pages. 1988.

Vol. 1309: B. Fiedler, Global Bifurcation of Periodic Solutions with Symmetry. VIII, 144 pages. 1988.

Vol. 1310: O.A. Laudal, G. Pfister, Local Moduli and Singularities. V, 117 pages. 1988.

Vol. 1311: A. Holme, R. Speiser (Eds.), Algebraic Geometry, Sundance 1986. Proceedings. VI, 320 pages. 1988.

Vol. 1312: N.A. Shirokov, Analytic Functions Smooth up to the Boundary. III, 213 pages. 1988.

Vol. 1313: F. Colonius, Optimal Periodic Control. VI, 177 pages. 1988.

Vol. 1314: A. Futaki, Kähler-Einstein Metrics and Integral Invariants. IV, 140 pages. 1988.

Vol. 1315: R.A. McCoy, I. Ntantu, Topological Properties of Spaces of Continuous Functions. IV, 124 pages. 1988.

Vol. 1316: H. Korezlioglu, A.S. Ustunel (Eds.), Stochastic Analysis and Related Topics. Proceedings, 1986. V, 371 pages. 1988.

Vol. 1317: J. Lindenstrauss, V.D. Milman (Eds.), Geometric Aspects of Functional Analysis. Seminar, 1986–87. VII, 289 pages. 1988.

Vol. 1318: Y. Felix (Ed.), Algebraic Topology – Rational Homotopy. Proceedings, 1986. VIII, 245 pages. 1988

Vol. 1319: M. Vuorinen, Conformal Geometry and Quasiregular Mappings. XIX, 209 pages. 1988.

Vol. 1320: H. Jürgensen, G. Lallement, H.J. Weinert (Eds.), Sem groups, Theory and Applications. Proceedings, 1986. X, 416 page 1988.

Vol. 1321: J. Azéma, P.A. Meyer, M. Yor (Eds.), Séminaire Probabilités XXII. Proceedings. IV, 600 pages. 1988.

Vol. 1322: M. Métivier, S. Watanabe (Eds.), Stochastic Analysi Proceedings, 1987. VII, 197 pages. 1988.

Vol. 1323: D.R. Anderson, H.J. Munkholm, Boundedly Controll Topology. XII, 309 pages. 1988.

Vol. 1324: F. Cardoso, D.G. de Figueiredo, R. Iório, O. Lopes (Eds Partial Differential Equations. Proceedings, 1986. VIII, 433 page 1988.

Vol. 1325: A. Truman, I.M. Davies (Eds.), Stochastic Mechanics a Stochastic Processes. Proceedings, 1986. V, 220 pages. 1988.

Vol. 1326: P.S. Landweber (Ed.), Elliptic Curves and Modular Forms Algebraic Topology. Proceedings, 1986. V, 224 pages. 1988.

Vol. 1327: W. Bruns, U. Vetter, Determinantal Rings. VII,236 page 1988.

Vol. 1328: J.L. Bueso, P. Jara, B. Torrecillas (Eds.), Ring Thec Proceedings, 1986. IX, 331 pages. 1988.

Vol. 1329: M. Alfaro, J.S. Dehesa, F.J. Marcellan, J.L. Rubio Francia, J. Vinuesa (Eds.): Orthogonal Polynomials and their Appli tions. Proceedings, 1986. XV, 334 pages. 1988.

Vol. 1330: A. Ambrosetti, F. Gori, R. Lucchetti (Eds.), Mathemati Economics. Montecatini Terme 1986. Seminar. VII, 137 pages. 198

Vol. 1331: R. Bamón, R. Labarca, J. Palis Jr. (Eds.), Dynami Systems, Valparaiso 1986. Proceedings. VI, 250 pages. 1988.

Vol. 1332: E. Odell, H. Rosenthal (Eds.), Functional Analysis. P ceedings, 1986–87. V, 202 pages. 1988.

Vol. 1333: A.S. Kechris, D.A. Martin, J.R. Steel (Eds.), Cabal Semi 81–85. Proceedings, 1981–85. V, 224 pages. 1988.

Vol. 1334: Yu.G. Borisovich, Yu. E. Gliklikh (Eds.), Global Anal – Studies and Applications III. V, 331 pages. 1988.

Vol. 1335: F. Guillén, V. Navarro Aznar, P. Pascual-Gainza, F. Pue Hyperrésolutions cubiques et descente cohomologique. XII, 1 pages. 1988.

Vol. 1336: B. Helffer, Semi-Classical Analysis for the Schrödin Operator and Applications. V, 107 pages. 1988.

Vol. 1337: E. Sernesi (Ed.), Theory of Moduli. Seminar, 1985. VIII, 2 pages. 1988.

Vol. 1338: A.B. Mingarelli, S.G. Halvorsen, Non-Oscillation Doma of Differential Equations with Two Parameters. XI, 109 pages. 1988.

Vol. 1339: T. Sunada (Ed.), Geometry and Analysis of Manifol Procedings, 1987. IX, 277 pages. 1988.

Vol. 1340: S. Hildebrandt, D.S. Kinderlehrer, M. Miranda (Ed Calculus of Variations and Partial Differential Equations. Proceedin 1986. IX, 301 pages. 1988.

Vol. 1341: M. Dauge, Elliptic Boundary Value Problems on Corr Domains. VIII, 259 pages. 1988.

Vol. 1342: J.C. Alexander (Ed.), Dynamical Systems. Proceedin 1986–87. VIII, 726 pages. 1988.

Vol. 1343: H. Ulrich, Fixed Point Theory of Parametrized Equivari Maps. VII, 147 pages. 1988.

Vol. 1344: J. Král, J. Lukeš, J. Netuka, J. Veselý (Eds.), Poten Theory – Surveys and Problems. Proceedings, 1987. VIII, 271 pag 1988.

Vol. 1345: X. Gomez-Mont, J. Seade, A. Verjovski (Eds.), Holomorp Dynamics. Proceedings, 1986. VII, 321 pages. 1988.

Vol. 1346: O. Ya. Viro (Ed.), Topology and Geometry – Ro Seminar. XI, 581 pages. 1988.

Vol. 1347: C. Preston, Iterates of Piecewise Monotone Mappings an Interval. V, 166 pages. 1988.

Vol. 1348: F. Borceux (Ed.), Categorical Algebra and its Applicatio Proceedings, 1987. VIII, 375 pages. 1988.

Vol. 1349: E. Novak, Deterministic and Stochastic Error Bounds Numerical Analysis. V, 113 pages. 1988.